Galaxy AI Unlocked

Discover What Makes Your Samsung Device *Truly* Smart

Dylan Blake

All rights reserved. No segment of this book or any portion therein may be distributed, reproduced, or transmitted in whatsoever manner, including photocopying, recording, or any other means of transmission without the express written permission of the writer, except for brief citations in a book review.

Galaxy AI Unlocked: Discover What Makes Your Samsung Device *Truly* Smart

First edition. August 27, 2024.

ISBN: 9781304032270

Copyright © 2024 Dylan Blake

Written by Dylan Blake

Contents

Before Reading this Book i

Introduction v

Section 1: Getting Started

1: Unboxing Your Galaxy Device 1

    Unboxing the Samsung Galaxy S24 Lineup 4

    Maintaining Your Device's Water resistance.................. 6

2: Power on and Setup your Device 9

    Setting up Your Galaxy Smartphone 9

    Transfer Data from Your Old Devices............................ 13

    Remapping the Side Key Function 17

Section 2: Beyond Voice Commands – Bixby Hidden Gems

3: Getting Started With Bixby 23

    What Is Bixby?... 23

    Bixby Voice 24

    Optimizing Bixby Settings 31

4: Bixby in Your Daily Life 39

    Communication and Productivity 40

    Entertainment and Home Management 42

    Health and Fitness 47

    Travel and Navigation 49

    Quick Commands 50

5: Special Bixby Features 59

    Bixby Vision 59

    Bixby Modes and Routines 69

Section 3: Galaxy AI – Discover More Smart Features

6: Smart Communication 87

    Call Assist 87

    Interpreter 97

    Chat Assist 103

7: Smart Productivity 111

    Circle to Search 111

    Transcript Assist 116

    Note Assist 121

    Browsing Assist 129

8: Smart Personalization 133

Generative Edit 133

Generative Wallpapers 137

Sketch to Image 141

Some Tips and Tricks 145

About Author 147

Index 149

Before Reading this Book

Thank you for choosing to read my book! Writing it has been fun and a good learning experience. I'm confident you'll find it an exciting and informative read.

This book is for anyone who wants to know more about their Samsung device. It will be most helpful to readers who use the latest Samsung flagship phones or older models with One UI 6.1. Even if you have an older model from 2021 or earlier or a non-flagship device, there is still plenty of useful information for you in this book.

Our focus in this book is on the things that make your Samsung phones smart – the "Advanced Intelligence (AI)" features. Advanced Intelligence is a broad term that covers artificial intelligence, such as Bixby, and other forms of intelligence.

When you start reading this guide, you'll notice it is divided into three sections. The first part introduces you to the basics of your device: how to set it up, power it on and off, etc. The second has to do with Bixby and all its hidden gems. You'd be surprised by what Bixby can do!

The last section is about "every other thing that is smart." Each section and chapter is crafted so that one leads to the next, taking you deeper into what makes your device intelligent. I suggest you carefully read it page by page without skipping any section and follow along on your smartphone to try the instructions for the best learning experience.

Please remember that Samsung regularly updates its devices with new features and improvements. While I've done my best to provide the most up-to-date instructions, some features or settings might differ slightly depending on when you purchase this book. If you notice any differences, I encourage you to take it as a challenge to explore the updates and see what they do.

There are two more things to know, "the table of contents" and "the index."

The table of contents at the beginning of this book provides an overview of the main topics and the page numbers where they can be found. Using it lets you quickly locate the information you need and jump to specific areas that interest you.

The index at the end of this book lists all the essential keywords and concepts covered and the page numbers where they can be found. This feature is a powerful tool for finding specific information quickly and efficiently.

I highly encourage you to use these tools as you read this book. They are designed to help you maximize your reading experience and ensure you don't miss any vital information.

Please don't hesitate to reach out with any feedback or questions. Your thoughts are invaluable and will help me keep improving my work. You can contact me directly at DylanBlakeGuides@gmail.com.

Thank you for your time!

Best regards,

Dylan Blake.

Introduction

There are lots of smart devices out there with their own unique features. Samsung smartphones are no different. The One UI 6.1 introduced Advanced Intelligence (AI) into Samsung phones. As a Samsung user, I thought it's high time we explore these AI features, old and new, so you can see what you've been missing out on and know the exciting things your device can do.

It is well-known that many Samsung users are not fond of the Bixby feature on their phones, and they have good reasons for this. Bixby can be frustrating, especially when it misunderstands something you've said repeatedly. However, I find Bixby to be a helpful companion. Sometimes, I need help turning off my phone's screen, turning on the flashlight, or playing music when my hands are wet or occupied, and asking Bixby for help becomes much needed.

There are innovative ways to make your life easier with the new Galaxy AI features. It's like having a magic wand that makes the difficult things easy or even the easy ones a breeze. Some might argue that the new wave of AI in our society could make us lazy, as we no longer challenge our minds but allow AI to do even the simplest of tasks. However, I think it helps us focus on what matters to us. These features include Circle to Search, Generative Edit, Note Assist, Transcript Assist, Call Assist, and more. The best part is Bixby can help with some of these features.

After reading this guide, you'll be able to use the AI features of your Samsung device more effectively, enhancing your lifestyle, productivity, and overall digital experience. Let's discover what makes your Samsung device truly smart and magical.

Section 1: Getting Started

1: Unboxing Your Galaxy Device

Before exploring more sophisticated methods to maximize and make use of the AI features available on your Samsung devices, it's worth taking a closer look at the designs of the Galaxy S24 series smartphones. These models are particularly noteworthy for introducing new Galaxy AI functionalities with the One UI 6.1 update. The images below will detail the designs of the Galaxy S24, S24+, and S24 Ultra. If you use other Samsung models with One UI 6.1, you might find this section less relevant.

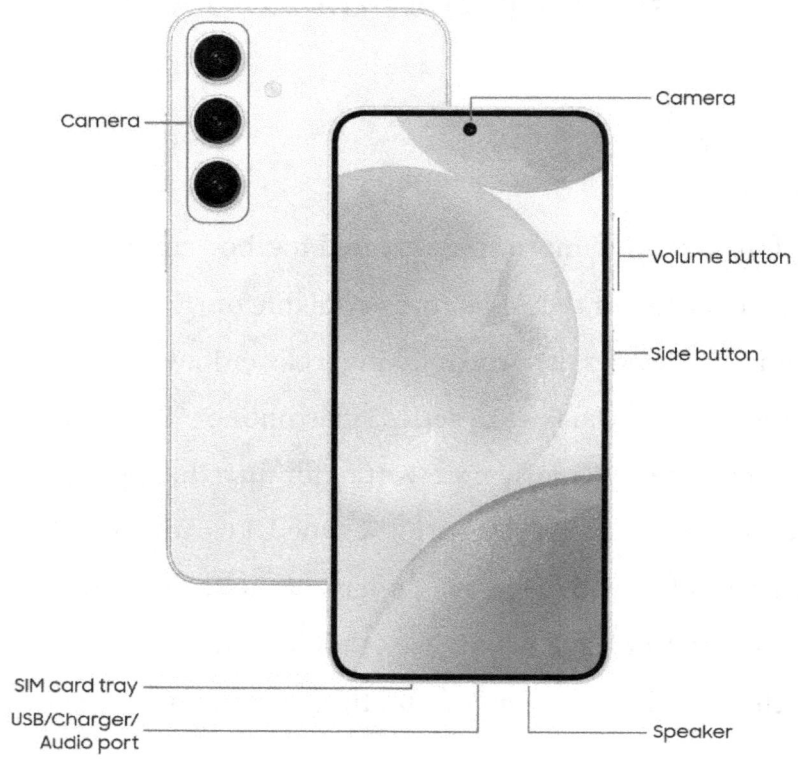

Samsung Galaxy S24

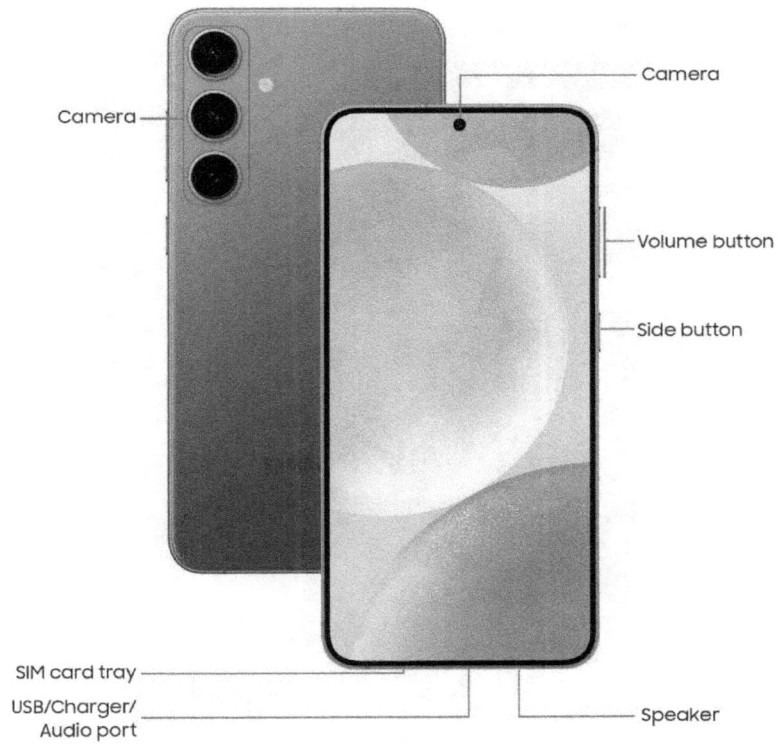

Samsung Galaxy S24+

4 Galaxy AI Unlocked

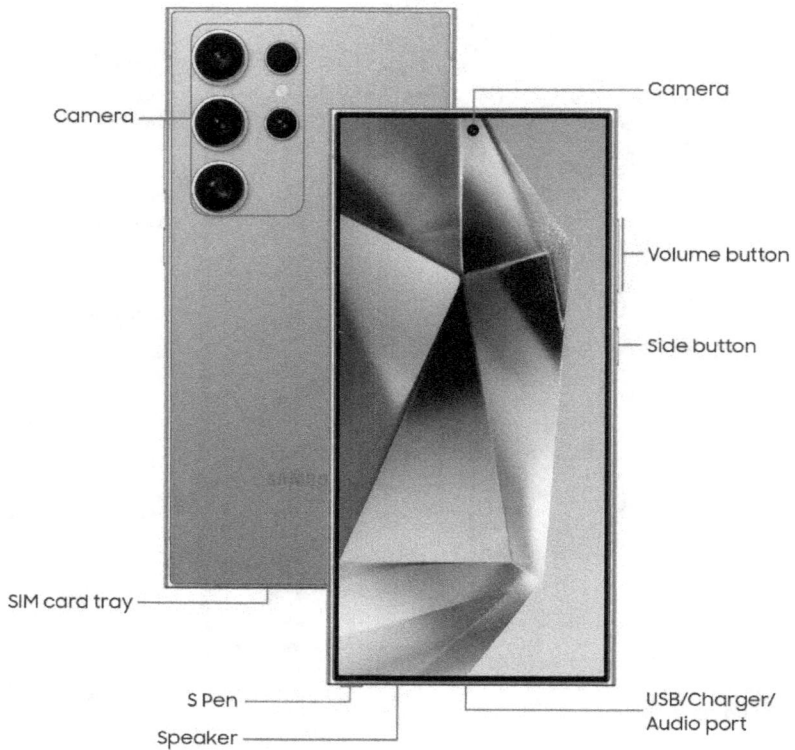

Samsung Galaxy S24 Ultra

Unboxing the Samsung Galaxy S24 Lineup

The Samsung Galaxy S24 smartphones come with everything you need to get started with your new device. Below is a list of things you would see in the box:

1. The Device
2. S Pen (Galaxy S24 Ultra)
3. USB-C Charging Cable

4. SIM Card Eject Tool
5. Quick Start Guide

It will help if you consider getting the following accessories for a better experience:

1. Samsung 45W USB-C Charger
2. A Glass Screen Protector (note that this may prevent the fingerprint scanner from working correctly).
3. A Case for Protection

The Galaxy S24 smartphones make use of a Nano-sim. It may also come with a preinstalled SIM called eSIM.

What is an eSIM?

It is a newly developed technology that allows you to switch between carriers easily and have more than one phone number on your mobile phone. An eSIM, "embedded subscriber identity module," is a small electronic chip implanted into a mobile device to serve the same function as the little plastic SIM cards.

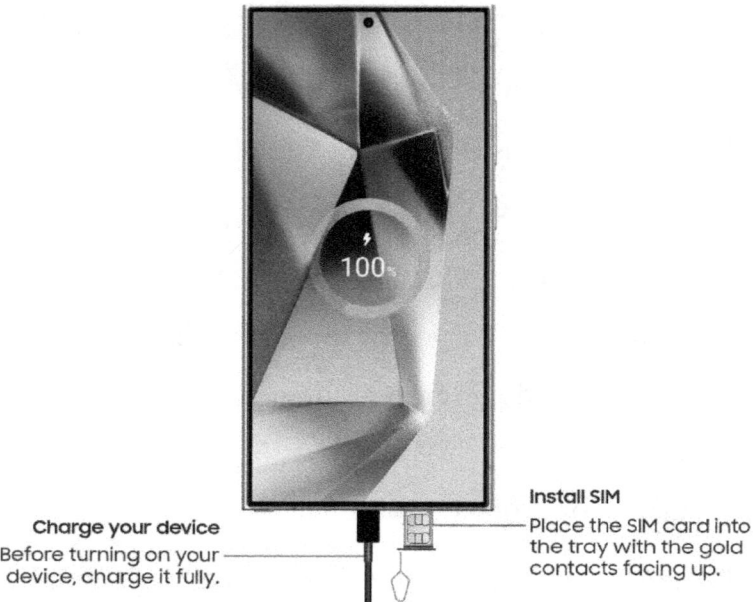

Charge your device
Before turning on your device, charge it fully.

Install SIM
Place the SIM card into the tray with the gold contacts facing up.

Maintaining Your Device's Water resistance

⚠️ **Caution:** To minimize the risk of electric shock and damage to your smartphone, do not charge your device when wet or in a damp environment.

It is essential to keep all compartments of your device closely sealed as it isn't impervious to water and dust in all states. For example, exposing your phone to fresh water without tightly sealing the SIM card tray may lead to damage.

Follow these guidelines carefully to prevent your device from getting damaged and maintain your phone's water and dust resistance.

- Your device has an IP68 rating. A test reveals that your phone can stay submerged in freshwater (more than 1.5 meters deep) for more than 30 minutes without damage. Always dry your device using a soft cloth after exposing it to freshwater. In cases where you expose your phone to liquids other than freshwater, rinse it with fresh water before drying it off.
- Seal or close any compartment that can be opened tightly to stop liquid from entering the system.
- Avoid exposing your device to water at high pressure, as it may lead to damage.
- Your touchscreen and other features may not respond well after your device is submerged in water. Also, the sound will be muffled when the speakers and microphone are wet, and you may not be heard clearly during a call. To fix this, thoroughly dry your device with a soft cloth. You can also expel water from the speaker and microphone by shaking your phone.

- The water resistance capability of your device can become compromised if your device is dropped.
- Your device may experience decreased sound or malfunction of certain features if dust or foreign objects enter the microphone, speaker, or receiver. And using a sharp object to remove the debris could damage your device and may alter its appearance.

Note: Other liquids may penetrate your device. Failing to rinse your device with fresh water quickly and drying it properly as instructed could result in operational and cosmetic issues.

2: Power on and Setup your Device

Use the side button (or side key) to turn on your device. Samsung recommends that you repair your device in cases where the body is broken or cracked before powering it on.

Setting up Your Galaxy Smartphone

When you first turn on your device, you may need to set up a few things to get started. The steps below will help you through the setup process for your Galaxy smartphone. You can skip any step you do not see on this list. It may be helpful to note that these steps vary depending on your region and carrier.

1. Press and hold the **side button** to Power on your device.
2. Select your preferred language at the "**Welcome!**" screen, then tap **Start** to go to the next stage.
3. In the "**For your review**" screen, tap **Agree to all (optional)**, then tap **Agree** (at the bottom right).

You can deselect **Sending of diagnostic data (optional)** if you wish.

4. On the "**Easy set up with another device**" screen, select an option: **Galaxy or Android device** (if you have an old Galaxy or Android device) or **iPhone or iPad** (if your previous device is an old iPhone or iPad), then follow the on-screen prompts. If you don't want to set up your device using your last smartphone, tap **Set up manually**.
5. Select a Wi-Fi network. You might be prompted to enter your password. When done, click **Connect** and then tap **Next**. After you've logged into your Wi-Fi network, your Samsung device automatically checks for updates and prepares your phone. You could also s**kip** and add the Wi-Fi networks after setup is complete.
6. Tap **Next** from the "**Copy apps & data**" screen, tap **Next** again, and follow the on-screen instructions to use Smart Switch to transfer data from your old device (Galaxy/Android or iPhone/iPad). Tap **Don't copy** to skip this section.
7. Sign in to your Google account, then hit **Next** to continue and follow the on-screen prompts. If you don't have a Google account, you can create one by

tapping **Create account**. You can also skip this process by tapping **Skip**. If you choose to skip, you can add your Google account after you finish the setup process.

8. From the "**Google Services**" screen, you can turn off **Use location**, **Allow scanning**, or **Send usage and diagnostic data** (most individuals prefer to turn off this option) depending on the services you find unnecessary. Tap **More** twice and then **Accept** to continue.

9. Select your preferred option from the "**Protect your phone**" screen, or you can do it later by tapping "**skip.**" You can set up your **Fingerprints** or **Face recognition**, but you must have a PIN, Password, or Pattern.

10. Insert your login details in the "**Samsung account**" screen to sign in to an existing account. To open a new account (or to skip this process), click "**Forgot password or don't have an account?**" Tap **Set up later in Settings** (at the bottom) and click **Skip**. You can add a Samsung account when the setup process is complete.

11. On the "**Samsung service legal information**" screen, tap **More,** select **All (optional),** and click **Agree**.
12. On the "**Samsung service permissions**" page, click **Agree** to gain access to all the services listed.
13. Select either Light or Dark as your preferred display mode in the "**Choose your display mode**" screen and tap **Next**.
14. On the "**Get recommended apps**" screen, unselect all of them (if possible and if you don't find them helpful) and click **Next.**
15. On the "**You're all set up!**" Screen, tap **Finish.**

Turn off and Restart your Device

Follow the steps below to turn your device on, power it off, and restart it.

- Press and hold the **side button** to turn on your device.
    - To power off your device, open the device notification panel by swiping down using two fingers from the top of the device screen and tapping the ⏻ **Power icon** (at the top right) > ⏻ **Power off.** Confirm your actions when prompted. You can also turn off your

device by simultaneously holding the side and volume down keys.

- To restart your device, open the device notification panel and tap the ⏻ **Power icon** > ↻ **Restart**. Confirm your actions when prompted. You can also reboot your device by simultaneously holding the side and volume down keys and then tapping ↻ **Restart**.

Transfer Data from Your Old Devices

Using the Samsung Smart Switch, you can transfer data from an old phone to your new Samsung Galaxy device if you didn't do so when you first turned on your device. The Samsung Smart Switch lets you smoothly move all the content in your old phone (such as contacts, messages, pictures, and videos) to your new device. This method can transfer data through a USB cable, Wi-Fi, or computer. We'll only look at how you can transfer data via a USB cable and Wi-Fi.

Note that when you use the Smart Switch app, you can only transfer content from your old phones to Samsung Galaxy devices.

Transferring Data via USB Cable

This method is recommended for large data transfer as it is relatively faster; however, you may want to charge both phones fully before starting this process because you'd be unable to charge them once you begin.

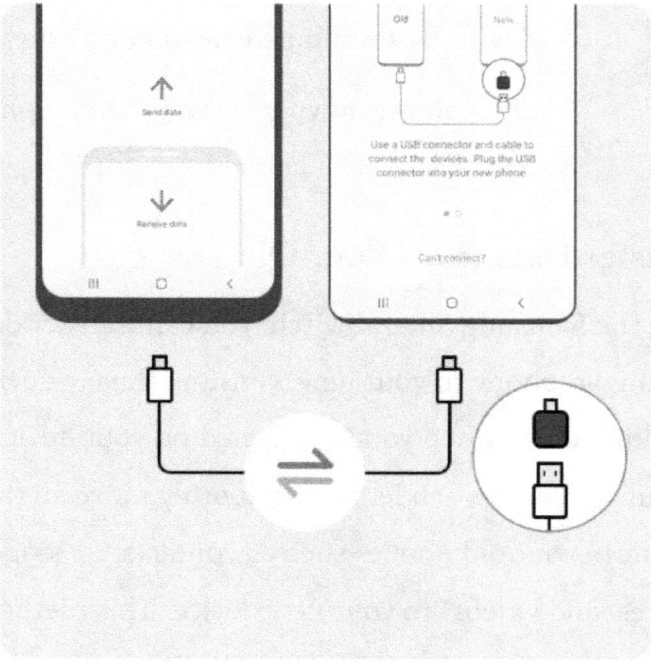

Note: A USB connector (i.e., a USB-OTG adapter) might be required to connect your new Galaxy device to your old phone. A USB-OTG adapter comes in the box with the Galaxy Z flip, Note10+, Note10+ 5G, S10+, S10, and S10e.

Follow the steps below to transfer content via a USB cable:

1. Connect both devices by using your old phone's USB cable. Most USB cables require an **adapter** to **connect your new device**.
2. Now, launch the **Smart Switch app** (make sure you have the most up-to-date version) on both devices. You can download and install it if you don't have it installed.
3. On the old device, tap **Send data** > **Cable;** on the new one, tap **Receive data** > **Galaxy/Android** or **iPhone/iPad.** Connect the old device to the new one with a cable. Smart Switch X-rays the old phone for transferrable data.
4. Choose the content you'd like to transfer to your new device. You'll see an approximate time for the transfer to be complete. If the transfer time exceeds an hour, it is advisable to go through the wireless transfer process if your devices don't have a full charge so that you can charge them during the transfer.
5. When you're ready to begin the process, tap **Transfer**.
6. At the end of the process, tap **Done** on the new Device and **Close** on the old one.

Transferring Data via Wi-Fi

Transferring data via Wireless becomes the best option for individuals with much content and phones that don't have a full charge. This method allows you to plug your devices into a charger during the transfer.

Follow the steps below to transfer content through Wi-Fi:

1. Install the Smart Switch app on both devices and plug both phones into their chargers while ensuring they are within 4 inches of each other.
2. When ready, launch the Smart Switch app on both phones. On the old device, tap **Send data;** on the new one, tap **Receive data,** and then select **Wireless** on both devices. You may have to choose the type of old device you're using on your new phone. Follow the on-screen directions, as some devices may require extra steps.
3. Complete the connection between both phones by tapping **Allow** on the old device.
4. Select the data you want to move on your new device and tap **Transfer** when ready to begin the process.
5. When the process is complete, tap **Done** on the new phone, and you're good to go.

Remapping the Side Key Function

You can assign a shortcut to the side button, such that it performs specific functions whenever you press it once, twice, or press and hold. The side button in recent Galaxy devices combines the power key and the Bixby key. The side button of your device supports four gestures, and we will look at each of them.

1. **Single Press:** This gesture is not customizable. It only functions to wake up or turn off the screen. You can also configure it to lock the screen instantly
2. **Double press:** Select the feature that launches whenever you double press the side button. To do this:
    - Go to **Settings** > **Advance features** and click **Side button.**
    - Click on **Double press** to enable (or turn off) this feature and select an option:
        - **Quick launch camera (default):** To open the camera app.
        - **Samsung Wallet quick access**: To swiftly access your Samsung Wallet.

- **Open app:** To open any app of your choice.

3. **Press and hold:** Select the feature that launches whenever you press and hold the side button (Note that you can't turn off this gesture). To do this:

    - Go to **Settings**, and click ⚙ **Advance features** > **Side button.**
    - Under **Press and hold,** tap to select an option:
        - **Wake Bixby (default):** This option is, by default, the feature that gets launched whenever you press and hold the side key.
        - **Power off menu:** It is best to select this option if you're not a heavy Bixby user so that you can quickly turn off your device by pressing and holding the side key.

4. **Press and hold the side and volume keys:** The side button performs other functions when held simultaneously with the volume down or up keys. Note that this feature isn't customizable, and each combination of the side and volume keys has a specific function.

- **Press and hold the side and volume-up keys:** When turned off, you can reboot your Samsung Galaxy device into recovery mode by simultaneously pressing and holding the side and volume-up keys.
- **Press and hold the side and volume-down keys:**
    - You can simultaneously press and hold the side and volume-down keys for two seconds or less to take a screenshot.
    - You can power off your device by pressing and holding the side and volume-down keys for over seven seconds. This feature also works even when your device is not responsive.
    - To get to the Power off menu, press and hold the side and volume-down keys for a few seconds — greater than two but less than seven seconds.

Section 2: Beyond Voice Commands – Bixby Hidden Gems

3: Getting Started with Bixby

What Is Bixby?

You may have heard some Samsung users repeatedly saying "Hi, Bixby," like they're having some weird conversation with their phone. And you'd be right! However, some users don't take advantage of this feature. Perhaps they're unaware of the remarkable things they can do with Bixby. If you're in this category, prepare to be amazed.

Bixby is a virtual assistant that is your gateway to a voice-activated universe. A simple "Hi, Bixby" awakens your device, transforming it into an intuitive partner that is always ready to assist. Bixby is an essential player in the Galaxy ecosystem, crafted to simplify your life by predicting your needs and offering intelligent recommendations. With Bixby, your voice has the commanding Power to get things done hands-free.

Bixby Voice

It is super easy to access Bixby on your Galaxy device by simply pressing and holding the side button just below the volume keys on the right side of your device (or the Bixby key by the left for older devices). After you've done that, you will be taken to the Bixby page; you should then carry these mandatory steps:

Note: If pressing and holding the side key doesn't bring up Bixby, this function may have been altered; *see* **Remapping the side key Function** (page 17) to change it back to Bixby. Alternatively, you can access Bixby by swiping up from the Home screen to open your Apps list, tapping the search box at the top, typing in "Bixby," and then selecting the Bixby app from the search results.

1. As stated earlier, pressing and holding the side key will give you an introductory message about Bixby. You may need to sign in to your Samsung account.
2. Choose your preferred language at the top right and tap **Start**.
3. Select your desired voice style by tapping a voice option to preview and then click **Continue**.
4. Review the required permissions and tap **Continue** to accept.

When your registration is complete, you've successfully activated Bixby! To interact with Bixby, either press and hold the side key when you're ready to talk or say "Hi, Bixby" if Bixby Voice wake-up is active.

Activating Bixby Voice wake-up

This feature allows Bixby to spring into action whenever you say "Hi, Bixby" or "Bixby." You can also fine-tune the Voice wake-up settings to suit your preference. Follow the simple steps below to get started:

1. To begin, swipe down from the top of any screen on your phone to reveal the Quick Panel. Then, tap ⚙ **Settings** in the top right corner to access your device's settings menu.

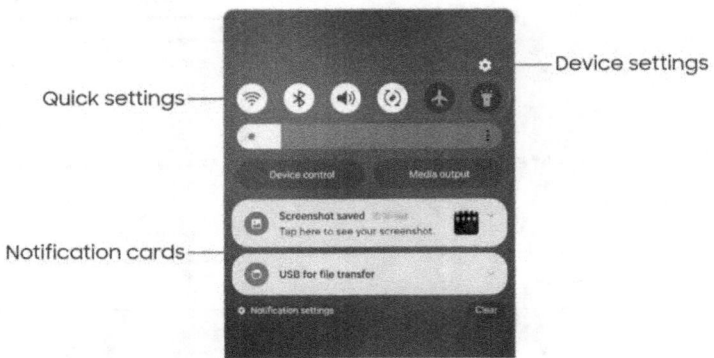

2. Next, scroll down and tap ⚙ **Advanced features** > **Bixby.**

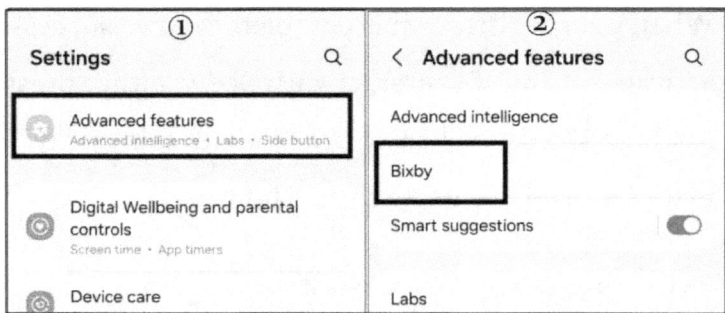

3. Click **Voice wake-up,** where you'll find other options you can customize to your preference. Tap the toggle button to turn on this feature.

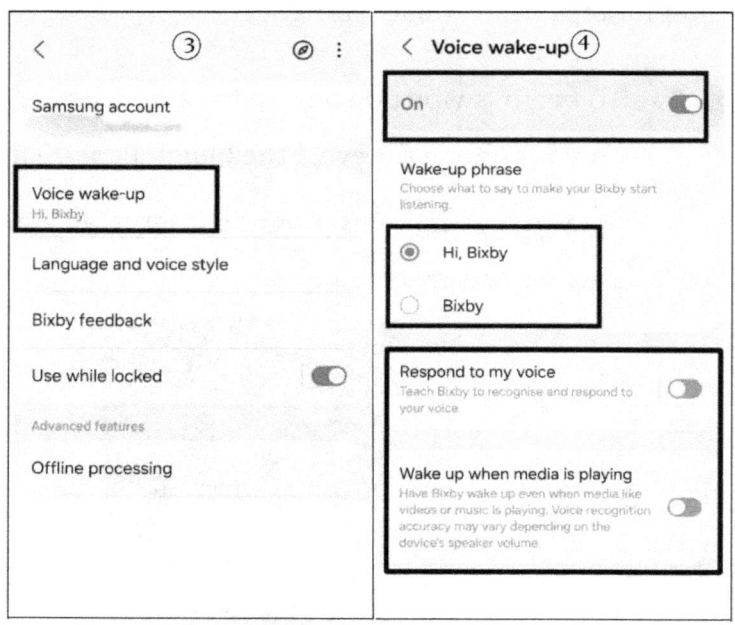

- In the "**Wake-up phrase**" section, you can choose "**Hi, Bixby**" as your activation phrase, ensuring that Bixby activates and listens to your commands only when you say this phrase. Alternatively, you can select "**Bixby**" as a shorter wake-up command. Opting for "Hi, Bixby" might be preferable to minimize accidental activations, as it requires a more deliberate phrase than merely saying "Bixby."

- o If you'd like Bixby to listen and respond to your commands even when music or a video is playing, activate the "**Wake up when media is playing**" option. To do this, tap the toggle button next to this feature. This setting ensures that Bixby listens for your voice even when a video is playing. Please note that the accuracy of voice recognition may diminish with higher speaker volumes.

Training Bixby to Recognize Your Voice

Simply activating Bixby Voice Wake-up might not be enough for Bixby to spring into action and carry out whatever command you have for it. However, for Bixby to respond consistently and accurately to your wake-up command, personalizing its voice recognition to your unique speech patterns (including your accent and pronunciation) is essential. This training focuses on ensuring that Bixby recognizes your voice when you say the wake-up phrase, such as "Hi, Bixby" or "Bixby," enhancing both the device's interactivity and security by preventing unintended activations by others.

Follow the steps below to enable this feature:

1. Go to **Settings** > ⚙ **Advanced features** > **Bixby** and tap **Voice wake-up**
2. Enable "**Respond to my voice**" by clicking the toggle button next to it.
3. Click **Start** on the "**Teach me to wake up to your voice**" page and follow the on-screen instructions, which involve repeating the words "Hi, Bixby" or

"Bixby," as the case may be, and as clearly as possible until your voice is registered five times.

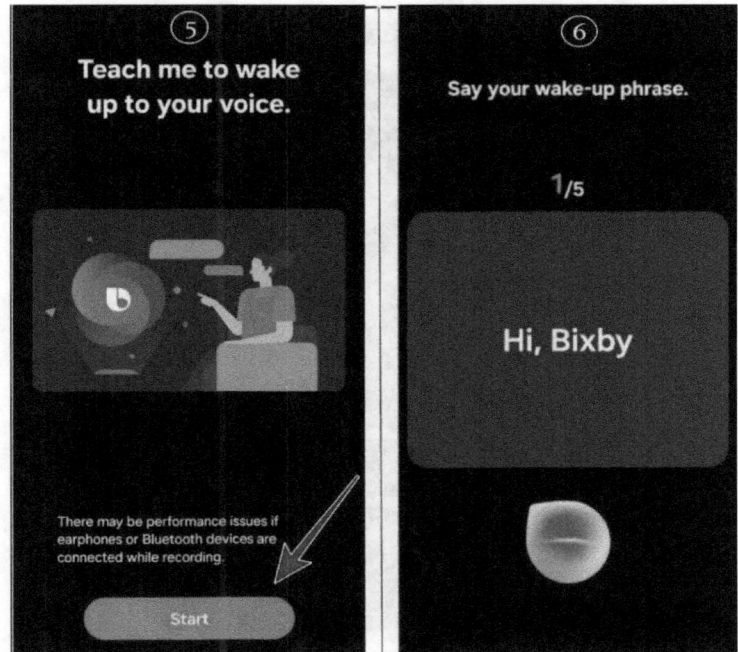

Improving Bixby Recognition Accuracy

After initially training Bixby to recognize your voice, you can do a few things to improve how well Bixby recognizes your voice.

1. **Retrain Bixby**: This process involves deleting your previous voice recordings to start afresh. The retraining should be conducted in a quiet environment, speaking slowly and clearly for optimal results. To delete your previous recordings,

navigate to **Settings** > ✻ **Advanced features** > **Bixby** > **Voice wake-up** > **Respond to my voice** and click "**Delete voice wake-up recordings.**"

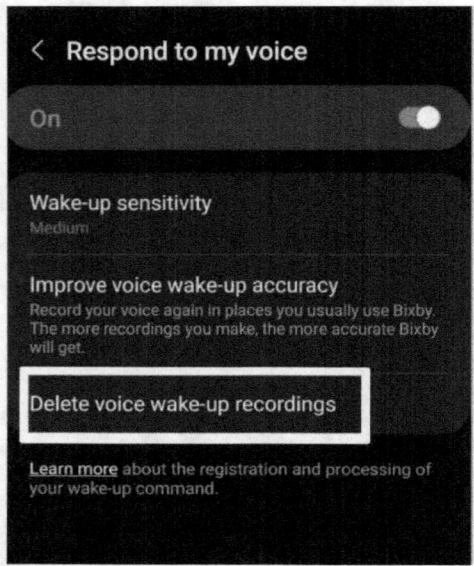

Tap **Delete** to confirm. Once your previous recordings are deleted, the "**Respond to my voice**" feature will be disabled, and to reactivate it, tap the toggle button ⬤ and follow the on-screen instructions to retrain Bixby with your voice.

2. **Adjust Wake-up sensitivity**: You can adjust Bixby's Wake-up sensitivity to your preference. The wake-up sensitivity setting controls how readily Bixby responds to your wake-up command. The default setting is medium, which is suitable for most users, but you can customize this to high or

low based on your environment and preferences. Navigate to **Settings** > ✼ **Advanced features** > **Bixby** > **Voice wake-up** > **Respond to my voice,** click "**Wake-up sensitivity,**" and select an option.

3. **Improve voice wake-up accuracy**: The "Improve voice wake-up accuracy" option allows you to enhance Bixby's ability to recognize your voice more accurately. This process involves recording your voice under various conditions where you frequently use Bixby, following steps similar to the initial voice training. This approach ensures that Bixby becomes even more attuned to your voice, enhancing its responsiveness and reliability in your daily use. Navigate to **Settings** > ✼ **Advanced features** > **Bixby** > **Voice wake-up** > **Respond to my voice,** click "**Improve voice wake-up accuracy,**" and follow the on-screen instructions.

Optimizing Bixby Settings

We'll look at Bixby settings and the ones to change for a better experience. Navigate to **Settings** > ✼ **Advanced**

features > Bixby. There are tons of fantastic adjustments you can make, and we'll be looking at some of them:

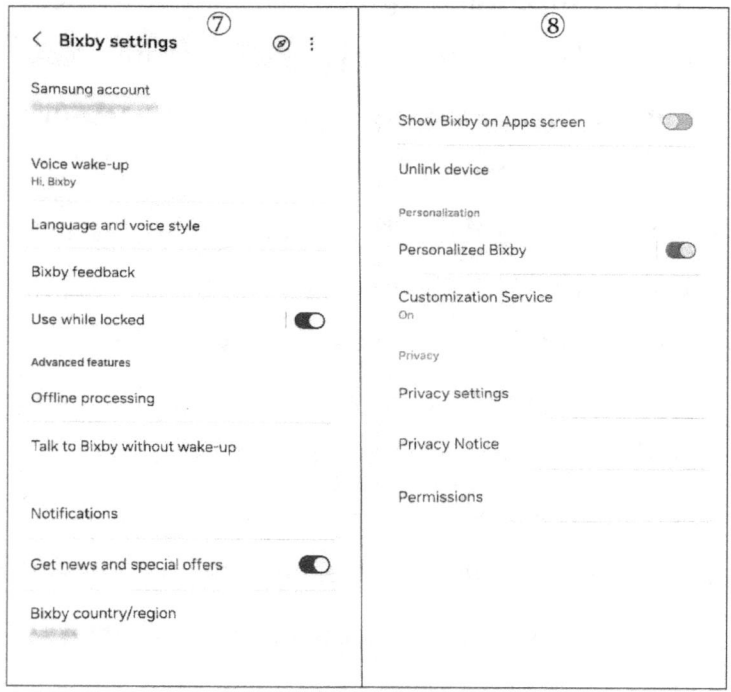

1. **Samsung account**: Here, you'll see your Samsung account profile with the option to sign out and edit other features.
2. **Voice wake-up:** You should have this feature Enabled already; if you don't, *see* **Activating Bixby Voice** in this chapter (page 25) to enable it. After enabling this feature, you can use Bixby by saying, "Bixby," or "Hi, Bixby." You can also

increase or reduce the Wake-up sensitivity, improve voice wake-up accuracy by re-recording your voice in areas where you frequently use Bixby, delete voice wake-up recordings, and enable Speak seamlessly.

3. **Language and voice style:** Change "Language" to the language of your choice and select your favorite **Voice style** by clicking on it. When you choose English (United States) as your preferred language, you have four voice options, numbered 1 to 4, available for selection. You may be prompted to download a voice whenever you select an option you've not previously downloaded.

4. **Bixby feedback**: Here, you can turn on or off "Sound feedback" and select your preferred "Voice response."
    - **Sound feedback**: I prefer it turned off, but you could leave it on so that whenever you say "Hi, Bixby," or "Bixby," you'll get a sound notification letting you know that Bixby is active and ready for your voice command.
    - **Voice response**: This setting lets you decide how Bixby responds. You have two options:

"**Always**" and "**Only with the wake-up phrase**." By selecting **Always**, Bixby gives voice feedback for all cases. However, I prefer the **Only with the wake-up phrase** option, which lets Bixby provide voice feedback when you use the wake-up phrase, when the phone cover is blocking the screen, when using navigation while driving, or when using a Bluetooth device or a headphone.

5. **Use while locked**: Enable this feature by tapping the toggle button to let Bixby perform tasks like making calls, sending messages, and checking the weather even while your device is locked. When you click this feature, you have two options to adjust to your preference:
    - **Allow personal results:** Having this off is preferable to prevent easy access to your personal information when an unwelcomed guest with a similar voice or a recording of you uses your wake-up phrase and asks Bixby to read your messages, check your last caller, and so on. However, if the issues

stated are of little or no concern to you, you can turn on this feature by tapping the toggle button next to it for faster access to your device.

- **Allow smart home control:** This feature lets you control SmartThings devices while your device is locked by saying your wake-up phrase and telling Bixby what home appliances you want to turn on or off. An unwelcomed guest can access your home appliances if they have a similar voice or a recording of your voice. You can turn on this feature by tapping the toggle button next to it.

6. **Offline processing**: I recommend turning this feature on. To activate Offline processing, tap the toggle button and follow the on-screen instructions, if any. This feature lets you execute some Bixby commands even while offline (i.e., not connected to the internet). For example, you can ask Bixby to take a screenshot, set a time, turn on the flashlight, or call Mom.

7. **Talk to Bixby without wake-up**: I recommend turning this feature on by tapping the toggle

button ⬤ and following the on-screen instructions, if any. You may need to download and install **Samsung on-device Resource (EN)**. This feature lets you say relevant commands to Bixby without using the Voice wake-up phrase (i.e., without saying "Hi, Bixby"). It allows you to accept and reject calls, dismiss or snooze Alarms, and dismiss or restart Timers by saying commands like "Answer phone" or "Reject call," "Dismiss (or snooze) alarm," and "Dismiss (or Restart timer)." This feature is available only in English (US) and Korean.

8. **Bixby country/region**: Change your Bixby Country/region to use capsules from that location.
9. **Show Bixby on Apps screen**: If you want Bixby to be visible on your apps list, you should activate this feature by tapping the toggle button ⬤ next to it. Doing so will allow you to access Bixby directly from your Apps screen.
10. **Unlink device**: Unlink a device you've recently signed into from your account.

11. **Personalized Bixby**: Customize the suggestions and results you get from Bixby based on your preferences.

4: Bixby in Your Daily Life

In this chapter, we will explore some simple voice commands that let Bixby take care of specific tasks for you. Picture Bixby as your buddy who is always ready to lend a hand by doing little tasks for you, answering essential questions, and helping you meet your personal goals. Integrating Bixby into your daily life makes tasks easier and seamless. We'll categorize these voice commands into four segments:

- Communication and Productivity
- Entertainment and Home Management
- Health and Fitness
- Travel and Navigation

We'll also introduce the concept of Quick commands – a way to execute multiple actions with a simple voice command. Then, we'll explore how, with SmartThings, a Quick command like "Good Morning" can turn on your smart TV, brighten your lights, and open your curtains.

Communication and Productivity

Bixby can significantly improve and enhance your communication and productivity. With its assistance, you can easily make calls, manage incoming calls, send, read, and delete a text message, and set reminders and alarms, among other features. We'll look at some examples you can try to see how to take advantage of them in your day-to-day activities.

Note: "Hi, Bixby" will be the wake-up phrase we'll use for all the examples in this book. Try experimenting with other instances you can think of to see what you can do with Bixby.

For calls, here are some things you can try:

1. "Hi, Bixby; call James."
2. "Hi, Bixby; call my most recent number."
3. "Hi, Bixby; delete the contact, James."

In the first example, Bixby will automatically call James. However, you may have to answer some follow-up questions if you have multiple contacts named "James." Let's say you have three contacts with James as their first name or surname; you may have to pick one when Bixby asks you which one to call by saying "The first one," "The second one," or "The third one."

In the second example, Bixby calls the last number in your **Recents** list, whether a missed call or the last number you dialed.

In the third example, Bixby will Bring up the contacts with the name James, and you may have to pick the exact one by saying "The first one" or "The fourth one" and then answering the follow-up question to delete the contact.

For messages and emails, here are some things you can try:

1. "Hi, Bixby; send a text to James."
2. "Hi, Bixby; read my last text."
3. "Hi, Bixby; delete the last text from James."
4. "Hi, Bixby; read my unread emails."

In the above examples, Bixby may ask you some follow-up questions to answer before the command you requested can be carried out. All you have to do is answer these questions as if a friend had asked them.

For alarms and reminders, here are some things you can try:

1. "Hi, Bixby, set an alarm for 7 AM tomorrow."
2. "Hi, Bixby; remind me to take out the trash by 8 AM."
3. "Hi, Bixby; set a timer for 1 hour."
4. "Hi, Bixby; delete all alarms (or reminders)."

Entertainment and Home Management

Bixby can be your go-to for fun and managing your home with just your voice. Do you want music to set the mood or a video for a quick laugh? Bixby's on it. How about a playful round of Rock Paper Scissors or a dice roll to decide your next move in a game? Just ask. And when it comes to settling down for the evening, you can let Bixby recommend a movie and even turn on your smart TV for you. Let's look at some practical examples.

For entertainment, here are some things you can try:

1. "Hi, Bixby; play a music by Lana Del Rey."
2. "Hi, Bixby; play a music video by Drake."
3. "Hi, Bixby; play Rock Paper Scissors."
4. "Hi, Bixby; roll a dice."
5. "Hi, Bixby; recommend a movie."

In the first example, Bixby plays a song by Lana Del Rey. You can also ask Bixby to "play a song;" that way, it randomly plays any music, or you can specify the song by its name or artist. You may need Spotify or some other music app for this to work.

In the second example, Bixby should open the YouTube app and play a random music video by Drake. You can also ask Bixby to play a stand-up comedy video. Try other variations to see what other things you can do.

In the third example, engage in a game of Rock Paper Scissors with Bixby. For the fourth, if you need to roll a dice for a board game and don't have one handy, Bixby can roll a virtual dice for you.

For the fifth example, Bixby recommends a movie for you to watch. You could also specify the genre you want. You'll need to install Netflix to use this function.

For Home management, here are some things you can try:

1. "Hi, Bixby; turn on the living room lights."
2. "Hi, Bixby; start the dishwasher."
3. "Hi, Bixby; lock the front door."
4. "Hi, Bixby; increase the air conditioner temperature by 5 degrees."

Note: None of the above examples will work until you have **SmartThings** set up, an app you can use to control all the smart devices in your home.

SmartThings

Let's briefly look at the Samsung app called SmartThings. This app lets you pair your smart home devices with your Samsung phone, such as the air conditioner, dishwasher, light bulb, robot vacuum, sensors, TV, etc., so you can easily control them using your smartphone.

The SmartThings app typically comes preinstalled on Samsung phones. You can easily download it from the Galaxy Store or Play Store if it's missing from your apps list. Just search for "SmartThings" and hit **Install**.

Note: Make sure the app is current by going to the Galaxy Store or Play Store, searching for **SmartThings,** and tapping **Update**. If it shows "**Open**" instead, then the app is up-to-date.

Adding Your Smart Home Devices

The steps below were written based on the most recent update at the time and may have little variations from yours.

1. Go to your apps list by swiping up from the Home screen. Find and tap on **SmartThings**.

2. You may need to agree to Samsung's terms and conditions, and SmartThings may require you to grant access to Wi-Fi and Bluetooth.

3. After you've opened the app, tap **Devices** at the bottom of the page.

4. To add your smart home devices, tap the plus sign at the top or click **Add device**.

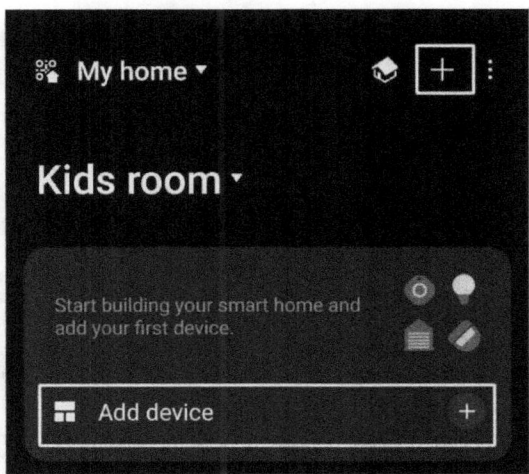

5. Do you have any Samsung smart devices like Refrigerators, TVs, robot cleaners, etc.? Then tap the **Add** button in the **Samsung devices** section, select what kind of device you want to add, and follow the on-screen instructions.

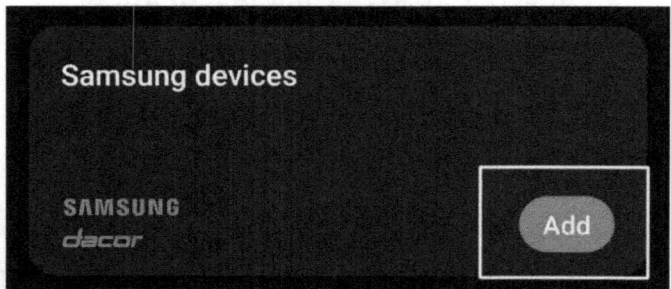

- Some devices allow you to connect by scanning a QR code on the device. Look for the one with the SmartThings label when there are multiple QR codes.

- You can connect directly using Wi-Fi and Bluetooth when you click "**Add without QR code.**"
6. You can connect with other partner devices by tapping the **Add** button on the **Partner devices** section and searching for a particular brand. Note that a SmartThings Hub (sold separately) may be required for some brands.

Note: After adding your devices, you can control them with Bixby Voice or the SmartThings app. Open the app and tap on **Devices** at the bottom of the screen. You'll see all your connected devices listed there.

Health and Fitness

Regarding your health and fitness, Bixby can come in handy to help improve your lifestyle and make living healthily easier. You can track your fitness, start an exercise, know the amount of calories in a food, and so much more. Let's look at some practical examples:

1. "Hi, Bixby; how many steps did I take today?"
2. "Hi, Bixby; how many calories are in a banana?"
3. "Hi, Bixby; start my running workout."
4. "Hi, Bixby; I had chicken for lunch."
5. "Hi, Bixby; remind me to drink water every two hours."

In the first example, Bixby will tell you the number of steps you've walked for the day. You may be required to set up the Samsung Health app if you've not already done so.

In the second example, Bixby tells you the number of calories in a banana. For the third, Bixby will track the duration and distance of your running workout. The Samsung Health app is also required for this to work. You can also ask Bixby to start other workouts like bicycling, walking, football (soccer), yoga, etc.

In the fourth example, Bixby helps log the meal you ate in the Samsung Health app, where you can further select other specifics like the part of the chicken and the portion size.

The fifth example also fits under Entertainment and Productivity, which concerns setting a reminder. Bixby sets a reminder that helps alert you to drink water every two hours.

Travel and Navigation

Did you know that Bixby can make navigation easier and travel more exciting? With your voice, you can save your parking spot, find places near you, set location-based reminders, and much more! Let's look at some examples:

1. "Hi, Bixby: save this location as my parking spot."
2. "Hi, Bixby: navigate me to the nearest coffee shop."
3. "Hi, Bixby: remind me to call Joe when I get Home."
4. "Hi, Bixby: what's the weather like in Paris today?"
5. "Hi, Bixby: how do you say 'where is the bathroom?' in Spanish?"

In the first example, Bixby will save your current parking spot. You may be asked some follow-up questions, which you must answer. To locate your parking spot, ask Bixby, "Where did I park?" and it should take you there.

In the second example, you'll be directed to the nearest coffee shop. You can also find the nearest restaurants, hotels, hospitals, tourist attractions, etc.

For the third example, you'll be reminded to call Joe once you get home. For this to work, you must have a location set as home. So when next you're home, say: "Hi, Bixby, set my current location as home." You can also set a reminder for when you leave home.

In the fourth example, Bixby tells you about the weather in Paris. For the fifth and last example, Bixby will translate the sentence for you into Spanish. You may be required to choose a translation service.

Note: Experiment with various queries to see what Bixby can do. It may sometimes misunderstand you or give irrelevant responses, so focus on simple, straightforward commands and stick with the queries you find most helpful for your everyday needs.

Quick Commands

Aside from asking Bixby straightforward commands that will carry out one specific action, you can create a Quick Command encompassing various actions. This way, Bixby will run multiple actions whenever you say a word or phrase.

For example, you can say: "Hi, Bixby, Good Night," and Bixby will turn on Do Not Disturb, disable Always On

Display (AOD), set the wake-up alarm for 8 AM, and carry out other commands. In this example, "Good Night" is the quick command, and the tasks Bixby does after are the actions that fall under the command.

The steps below will show you how to access Quick Commands.

1. To begin, swipe down from the top of any screen on your phone to reveal the Quick Settings panel. Then, tap ⚙ **Settings** in the top right corner to access your device's settings menu.

2. Navigate to ⚙ **Advanced features** > **Bixby** and click the three vertical dots at the top right.

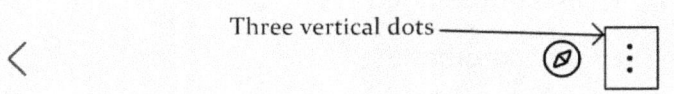

3. Next, click **Quick commands.**

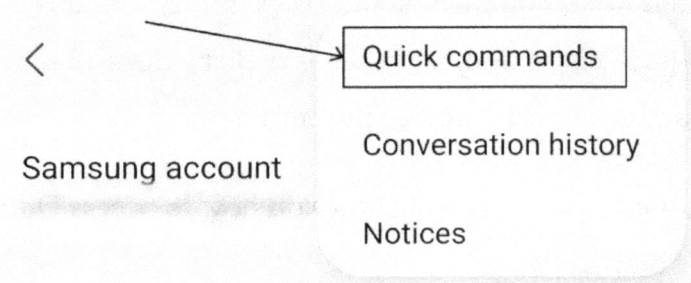

Create and Use Quick Commands

When you navigate to Quick Commands, you have two options at the base of your screen.

- Recommended
- My commands

Recommended

When you tap **Recommended**, you have a list of preset quick commands that perform multiple actions when you say the word or phrase assigned to them. You can also customize them to suit your needs better. We'll use the Quick Commands that fall into "**Daily Routine**" as an example.

1. In the **Recommended** tab, tap the "**Daily Routine**" capsule.

2. On the Daily Routine page, you'll see various options. Find and click on the "**Good Night**" capsule.

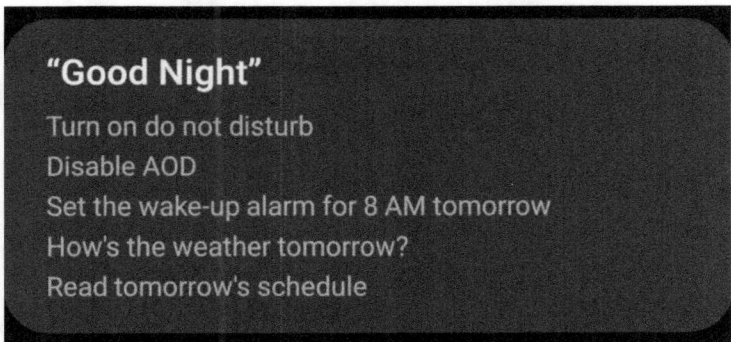

3. Click **Save** at the bottom to start using this phrase. Tap **Edit** to rename what you say, remove some commands, or add a command.

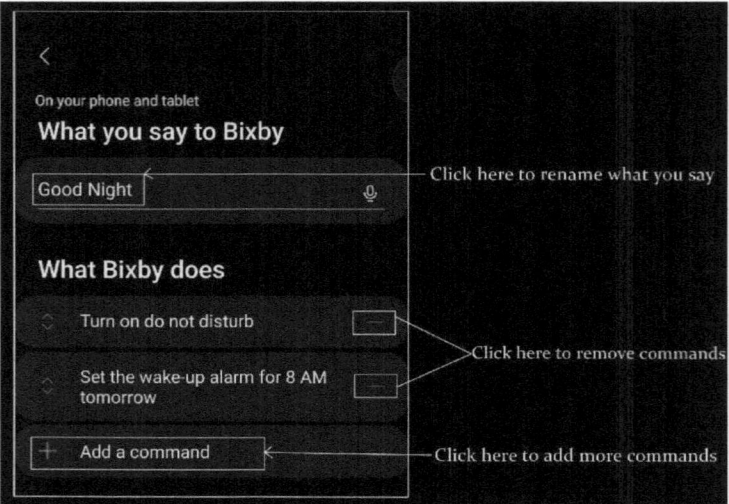

You can edit a command by clicking on it. Let's say your wake-up time is 6 AM, and if you want to edit it to that, click on the command "**Set the wake-up alarm for 8 AM tomorrow** (*see* last image)" and manually edit the "**8**" to "**6**."

4. While in edit mode, you can toggle on "**Run quickly**" so the commands run faster by skipping

the Bixby voice response. Tap **Run** to start the quick command.

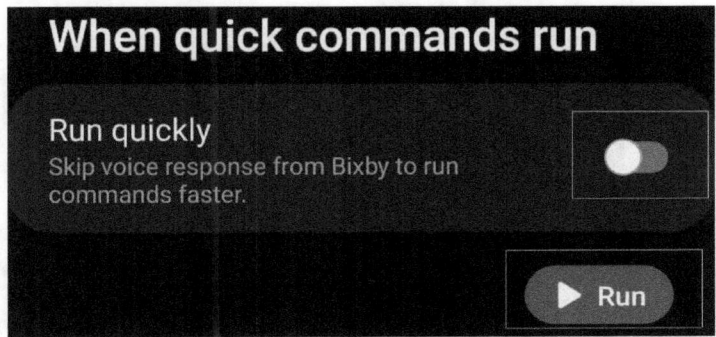

5. After clicking **Save**, you'll see this command in the "**My commands**" tab.
6. To use this command, say, "Hi, Bixby, Good night."
7. Create a "**Good Morning**" quick command by following the above steps, but this time, find and click on the "**Good Morning**" capsule instead of the "**Good Night**." We'll use these quick commands later in this book (*see* **Using Bixby Routines** (page 70)).

My Commands

When you tap "**My commands**," you'll see a list of your Quick commands. Here, you can edit, share, delete your existing commands, or create new ones.

1. You can create a quick command on the **My commands** tabs by tapping the plus sign at the top right.

   o   After you click the plus sign, you have three segments: "**What to say to Bixby**," "**What Bixby does**," and "**When quick commands run**," similar to when you edit a preset quick command as explained earlier.
2. On the **My commands** tab, tap a quick command to edit, share, delete, or add to your phone's home screen.

Quick Commands and SmartThings

Yes, we know a single voice command can carry out multiple actions, but did you know you can include controlling your smart home devices in these actions? The steps below will walk you through the process. We'll use the "Good Night" preset Quick command we saved earlier.

4: Bixby in Your Daily Life   57

1. First, open Quick commands, but this time by saying, "Hi, Bixby, open Quick commands."

2. Tap **My commands** at the bottom to see your saved commands.

3. Click on the "**Good Night**" Quick command and tap **Edit** at the bottom.

4. Click on **Add a command**.

5. You can type in the command, for example, "Turn off TV," then click **Done** or tap the microphone icon and say "Turn off TV" instead of typing it.

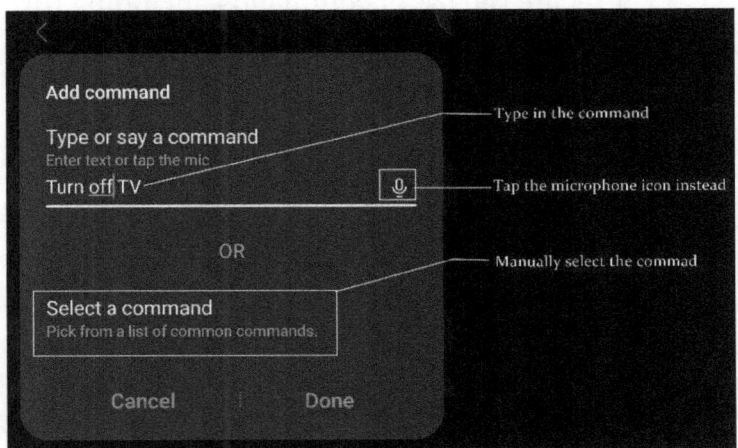

You can also manually select a command when you tap "**Select a command**" and follow these steps:

- On the page that appears, scroll down and click **SmartThings**, or you can tap the search box and type in "SmartThings."

- Next, type "TV" into the search box and tap **Turn off the TV**.

Note: You must have added your smart TV to your SmartThings app for this function to work. *See* **Adding Your Smart Home Devices** (page 44).

6. After adding the "Turn off TV" command, tap **Save**. Now, whenever you are ready to go to bed, all you need do is say, "Hi, Bixby, Good Night." Bixby will execute the list of commands, including turning off your smart TV.

5: Special Bixby Features

Bixby has many features; we'll look at the ones I find unique and special. There is more to Bixby than voice commands; it's time we see what else Bixby can do. Below are the unique features:

1. Bixby Vision
2. Bixby Modes and Routines

Bixby Vision

This feature bridges the gap between the digital and physical worlds, making it possible to interact with your environment using your phone's camera. Bixby Vision is Samsung's version of Google Lens and can be helpful in certain situations.

Updating Bixby Vision

Before we go into more details, let's ensure the Bixby Vision app is up-to-date. Follow these steps to update it:

60   Galaxy AI Unlocked

1. Swipe up from your Home screen to go to your apps list, then tap on the search box at the top and type in "Galaxy Store."

2. Select the **Galaxy Store** app that appears to open it, and click on the search icon at the top right.

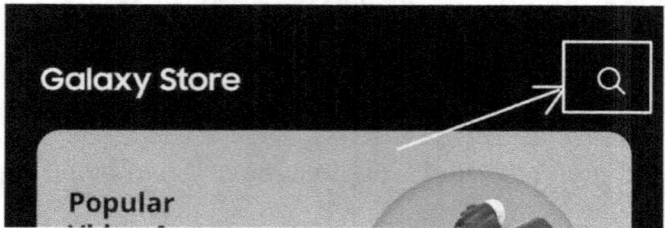

3. Type in "Bixby Vision" and click on the search icon in the bottom right of the keyboard. Tap the **Bixby**

Vision app and click **Update** at the bottom. If it shows **Open**, then the app is up-to-date.

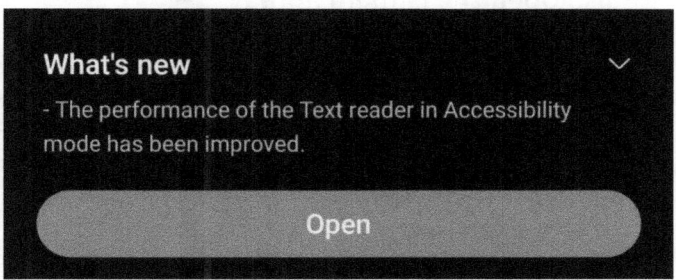

Opening the Bixby Vision app

There are two ways you can access the Bixby Vision app. One way is from your apps list, and the other is from the camera app.

Try this: "Hi, Bixby, open Bixby Vision."

- **Apps list**: From your Home screen, swipe up to access your apps list. You can find the Bixby Vision app by scrolling through the apps or using the search function at the top. In the search bar, type "Bixby Vision" and click on the app that appears.

- **Camera app**: To access Bixby Vision this way, you have to open the app by swiping up from the Home screen to access the app list and typing in "camera" in the search box at the top. Once in the camera app, swipe left to **More** and click **Bixby Vision** at the top. Double-pressing the side button should also open the camera app.

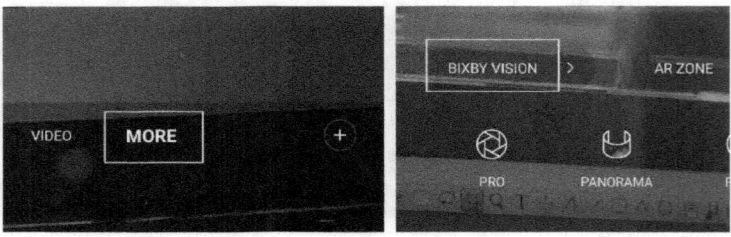

Bixby Vision Features

Let's look at what you can do with Bixby Vision once you open the app. You can play around with five options: Translate, Text, Discover, Wine, and Accessibility (activation may be required).

1. **TRANSLATE**: This option becomes useful when you see a text in a foreign language you need to understand. Let's imagine you're traveling in France, and you come across this sentence:

"Cette phrase a été écrite en français et traduite en anglais à l'aide de l'option Traduire de Bixby Vision."

If you want to know what it means, here's how to do it:

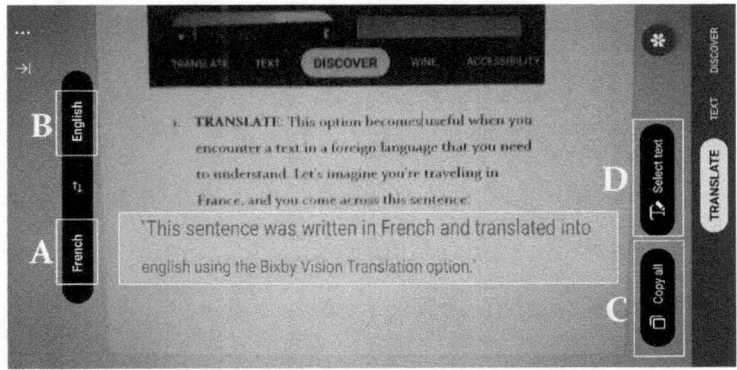

- o **Open Bixby Vision**: open the app through the camera app or from your app list and select **TRANSLATE**.
- o **Set Language**: Before pointing your camera at what you want to translate, set the source language to **French (A)** and the target language to **English (B)**. It is set to **Auto** by default, which automatically figures out the source language.
- o **Point to Translate**: Point your camera at the text you want to translate; your device should translate it in real-time. The

interface will display two more options: **"Copy all (C)"** to copy all text on your screen and **"Select text (D)"** to select the specific part of text on your screen.

Note: Tap anywhere on your screen to capture a still image for translation.

2. **TEXT**: This option becomes useful when you see a text you want to copy. Imagine reading a book and coming across a quote you'd like to save or send to a friend. Instead of typing it manually, you can use this feature. Here's how:

Quote by me: *"A picture is worth a thousand words, but your smile–a thousand emotions."*

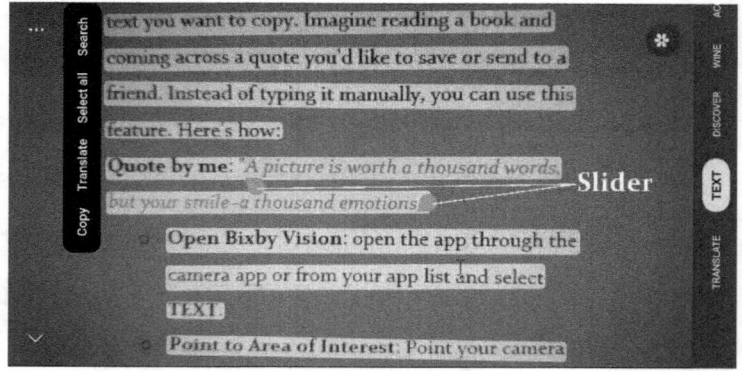

- **Open Bixby Vision**: open the app through the camera app or from your app list and select **TEXT**.

- **Point to Area of Interest**: Point your camera at the text you want to capture, and it will automatically capture and highlight the text.
- **Tap to Select**: Tap any word on the screen to select it. You can move the slider to select more words or tap "**Select all**" to select the entire text. You can **copy** the selected text, **translate** it into any language, and **search** for it online.

3. **DISCOVER**: This feature lets you point your camera at an object to find similar images on Pinterest. It can also attempt to identify celebrities and unique features of places and restaurants. Imagine you go camping and encounter an unknown insect; you can use this feature to identify it. Follow the steps below:

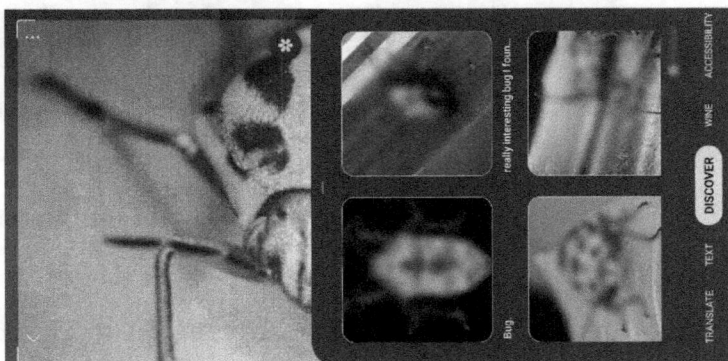

- **Open Bixby Vision**: open the app through the camera app or from your app list and select **DISCOVER**.
- **Point to Area of Interest**: Point your camera at the thing, animal, place, or person you want to identify, and it will automatically capture and show similar images.

Note: This feature isn't very accurate, but it can give you valuable results.

4. **WINE**: With this feature, you can get more information about a bottle of wine by scanning the label using Bixby Vision. Follow the steps below:

- **Open Bixby Vision**: open the app through the camera app or from your app list and select **WINE**.
- **Point to Area of Interest**: Aim your camera at the label on the wine bottle. Bixby Vision

will automatically capture the label and search for information about the wine.

- ○ **View Information:** If the search is successful, Bixby Vision will display details about the wine, including price, rating, acidity, and other characteristics. You can scroll up to see more information.

5. **ACCESSIBILITY**: This feature lets you use your smartphone camera to describe scenes, identify objects, read text aloud, and detect colors. It makes it easier for visually impaired users to interact with the world around them effectively. Follow the steps below to activate and use this feature.

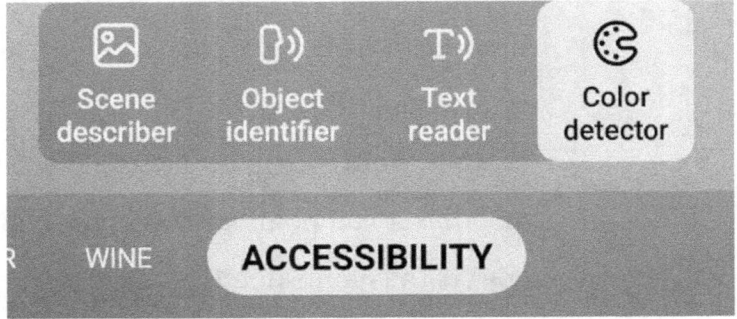

- ○ **Activate Accessibility Mode:** If you cannot find this feature on your Bixby Vision app, then you'll need to activate it by following the steps below:

- Open Bixby Vision through the camera app or from your apps list.
- Tap the three vertically stacked dots in the top right and tap **Settings**.

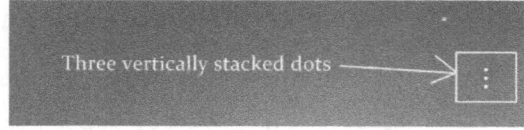

- Toggle on the switch for **Accessibility modes**.

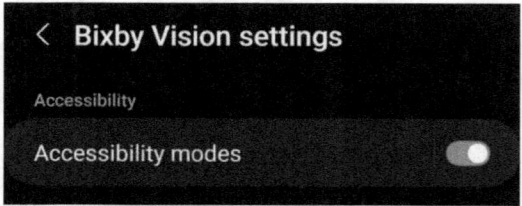

- o **Scene describer:** This feature analyzes and describes what you're looking at, including people, objects, and the surroundings.
- o **Object identifier:** This feature identifies any object you point your camera at and speaks the name aloud.
- o **Text reader:** This feature helps you read text. Point your camera at the text, and Bixby will read it aloud.
- o **Color detector:** This feature helps you to determine the actual color of things and can be helpful if you're color blind. Point your

camera at something, and Bixby will announce the color.

Bixby Modes and Routines

This feature uses the "IF" and "THEN" logic to automate actions on your device, allowing you to set up specific conditions (IF). When these conditions are met, your smartphone will automatically perform your specified actions (THEN). For example, you could set a routine to set your phone's sound mode to mute at night, switch to mobile data when Wi-Fi is weak, or turn on secure Wi-Fi when you leave home.

You can also set a routine using **Quick commands** (*See* page 50). Now, instead of saying "Hi, Bixby, Good Night," you can set a sleep routine such that when it's 11 PM (IF), your phone will execute the "Good Night" (THEN) Quick command.

Accessing Bixby Modes and Routines

There are two ways you can access the Modes and Routines app. One way is from your apps list, and the other is through your device's settings.

Try this: "Hi, Bixby, open Bixby Routines."

- **Apps list**: From your Home screen, swipe up to access your apps list. You can find the "Modes and Routines" app by scrolling through the apps or using the search bar at the top. Type "Modes and Routines" in the search bar and tap the app when it appears.

- **Device's settings**: Swipe down from the top of any screen on your phone to reveal the Quick Settings panel. Then, tap ⚙ **Settings** in the top right corner to access your device's settings menu. Next, scroll down and tap "**Modes and Routines**."

Using Bixby Routines

Let's look at how to create Bedtime and Wake-up time routines using Quick commands. With these examples, you should be able to generate your own routines. I'll also give some tips to make creating your routines a breeze.

Modes   Routines

1. **Bedtime routine**: First, tap **Routines** at the bottom right and click the plus sign at the top to create a new routine.

- Next, in the "IF" section, tap "**Add what will trigger this routine.**" Here, you'll choose the conditions that should be met for the routine to begin.

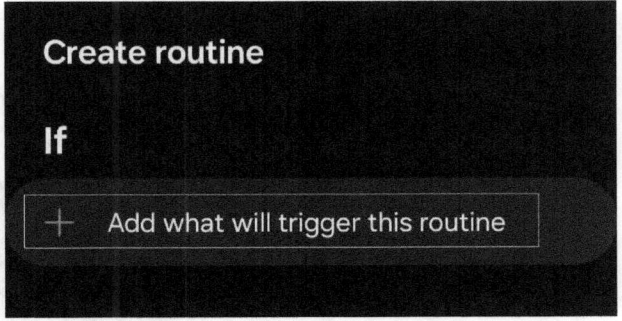

- Find and click on "**Specific time**" under the **Time** section. You can select "Time period" if you prefer to set how long you want the routine to last. You can also explore other options under Modes, Manual, Context, and others to see what more you can do.

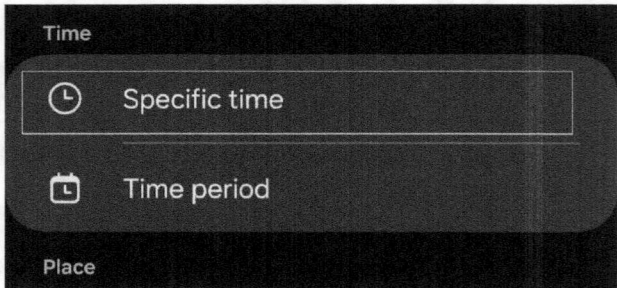

- On the next page, there are three capsules or sections:
    - Select "**Time**" in the first section. I find the "Sunrise" and "Sunset" options less accurate.

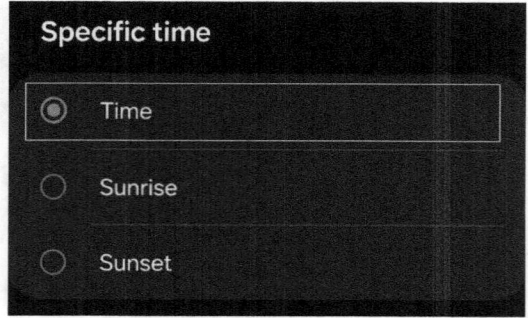

    - In the next section, set the time to **11 PM** (by swiping the numbers up or

5: Special Bixby Features   73

down) or choose a time that works best for you.

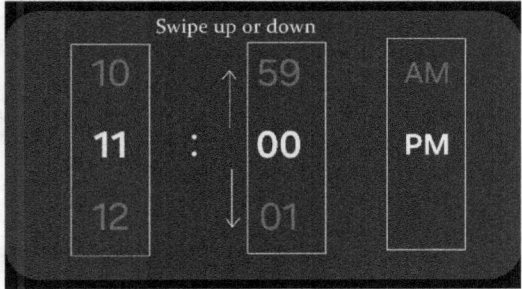

- In this last section, you can choose "Every day" if you want the routine to run daily, "Every week" if you're going to pick specific days of the week, or "Every month" to select specific days in a month.

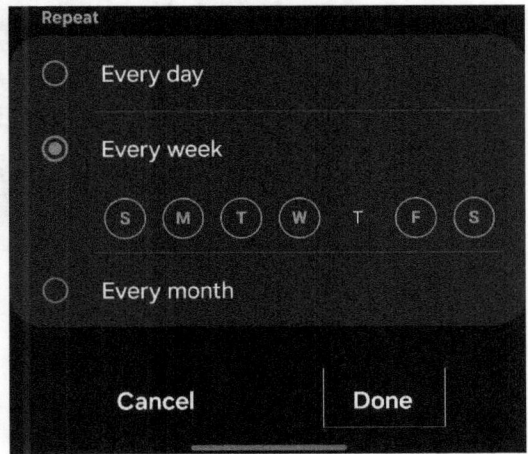

- Click **Done** at the bottom.

- Next, in the "Then" section, click on "**Add what this routine will do.**"

- Scroll down and click on "**Bixby voice**."

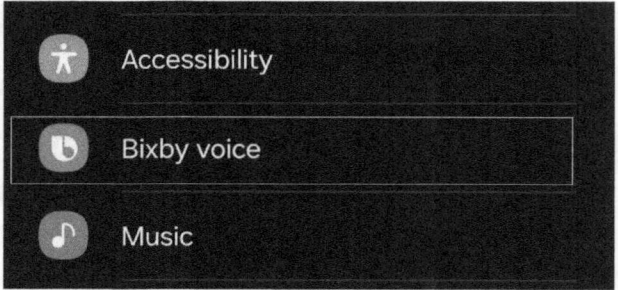

- Next, select "**Run quick command.**"

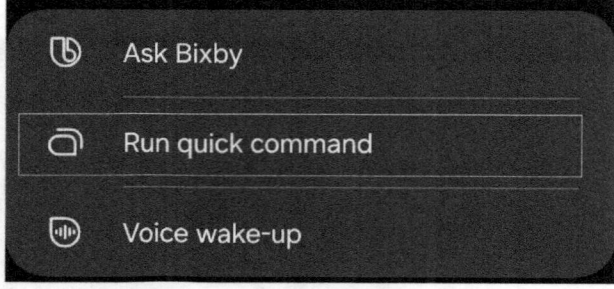

o Click the "**Good Night**" quick command and tap **Done** (*see* page 48 to create a quick command).

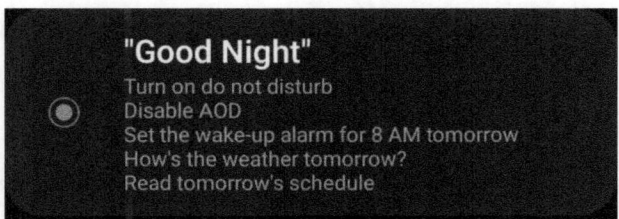

Note: You can expand this routine by adding more actions. Click **Add action** in the "Then" section, find and tap **Battery**, click **Power saving,** ensure it is set to **On**, and tap **Done,** which will turn on Power Saving mode when this routine is activated. Feel free to explore more options.

o Next, tap **Save** and edit the Routine name to "**Bedtime**." This action completes the setup process. By 11 PM daily (or whatever days of the week you selected), Bixby Routine will automatically run the "Good Night" quick command.

2. **Wake-up time routine**: Again, tap **Routines** at the bottom right and click the plus sign at the top to create a new routine. Follow the exact steps as in the Bedtime routine, but set the time to 8 AM or

choose a time that suits you. When you get to the "**Run quick command**" section, select the "**Good Morning**" quick command instead of the "**Good Night**" and you're good to go. This way, the Good Morning command will turn on the features turned off by the Good Night command.

> **"Good Morning"**
> Turn off do not disturb
> Enable Always On Display
> What is today's date?
> How's the weather?
> Any weather alerts?
> Read today's schedule

Note: Your routines will not function when your device is locked. You will need to unlock your device before the routine can proceed. However, by activating the **Used while locked** feature, Bixby can carry out your routines even while your device is locked (*See* page 34).

Using Bixby Modes

Modes are meant to help you focus on a specific task. For example, if you're driving, you can turn on Driving mode to help reduce distractions like calls and notifications. Similarly, when at work, you can turn on work mode to help maintain concentration. Let's take a closer look at Driving mode as an example.

Note: You can add a mode by scrolling to the bottom and tapping **Add mode**.

1. To turn on Driving mode, click on **Driving** and tap **Start**.

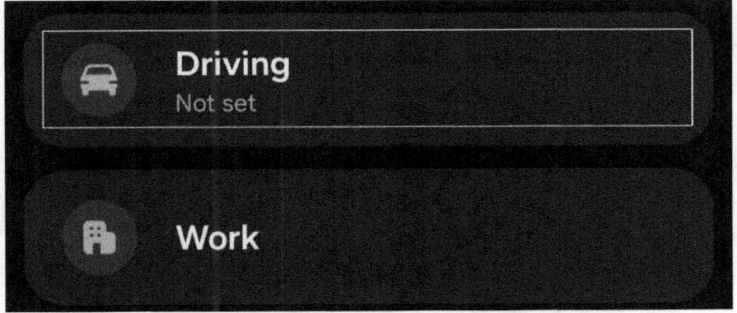

2. On the "**Set Driving mode to turn on automatically**" page, select **Android Auto** if your car supports it, or select **Bluetooth device** and choose a device. This setup ensures that "Driving mode" will automatically activate when your phone connects to the selected Bluetooth device or Android Auto. Tap **Next** to go to the next page.

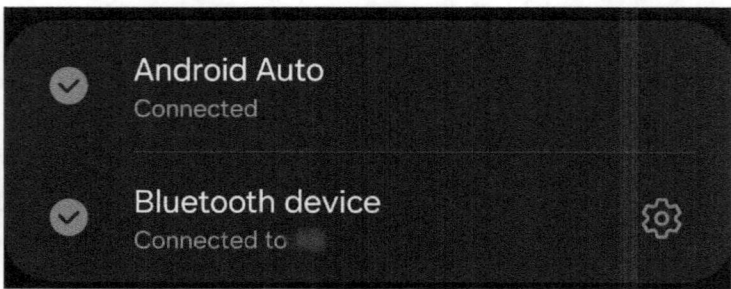

Note: **Android Auto** still requires a connection to your car's Bluetooth. Connecting to a **Bluetooth device** can be your car's or an external Bluetooth device.

3. Tap **Do not Disturb** to select it if you want to mute calls, messages, and notifications, or click **Skip** to go to the next page.

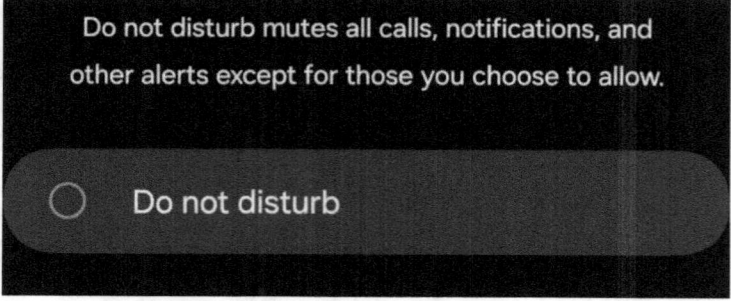

- After selecting **Do not Disturb**, tap **Calls and messages** to choose who can reach you when "Do not Disturb" is active.

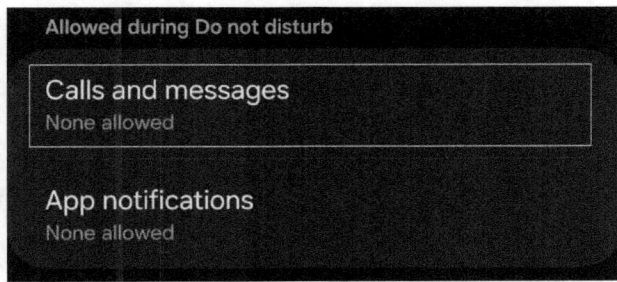

- In the "**Allow during Do not disturb**" section, tap **Add contacts** and select the contacts that can reach you.
- In the "**Also allow**" section, tap **Calls** or **Messages** and select an option (preferably "Favorite contacts only"). Toggle the switch for **Repeat callers** to allow anyone to reach you if they call twice within 15 minutes.
- Use the back button to return to the previous page.

o Tap **App notifications,** click **Add apps,** and select which apps can send notifications when "Do not Disturb is active." After choosing the apps, press the back button to return to the previous page.

o Click **Done** and tap **Next**.

4. You have four options on the "**Choose what to do in Driving mode**" page. We'll skip the first one; "**Open an app or do an app action.**" Feel free to explore this option later; it can be helpful, for example, to open Google Maps when Driving mode is activated.

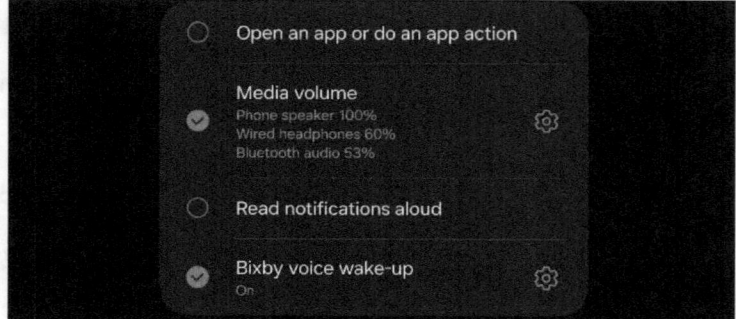

- Tap **Media volume** and adjust the volume for phone speakers and others to your preference, then tap **Done**. The next time Driving mode is activated, your phone's media volume will adjust to these settings.
- Click **Read notifications aloud** if you want your notifications to be read aloud while driving. You may need to grant **Modes and Routine** permission to access your notifications. Tap **Setting**, then find and toggle on **Modes and Routines** to grant this permission.

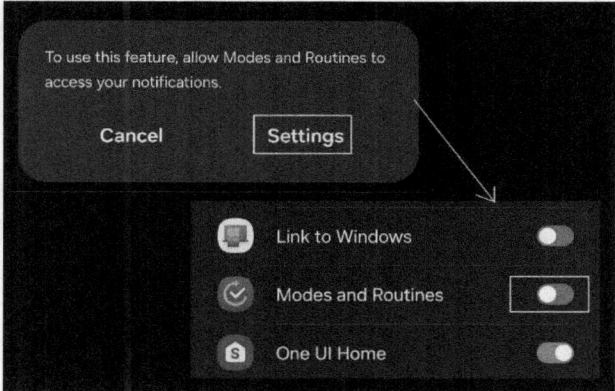

- Tap **Bixby voice wake-up**, select **On** (or **off** if you prefer), and click **Done**. By selecting **On**, Bixby will respond to your voice whenever you say the wake-up phrase (Hi, Bixby) followed by the command.

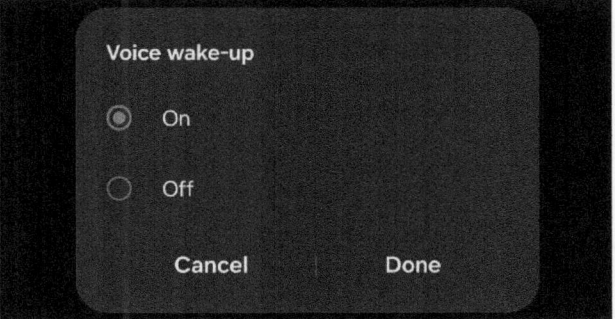

5. Tap **Done**.
6. **Driving mode** will be activated when next you're in your car and connect to the car's **Android Auto** or **Bluetooth device**. You can also manually turn it on by clicking **Turn on**.

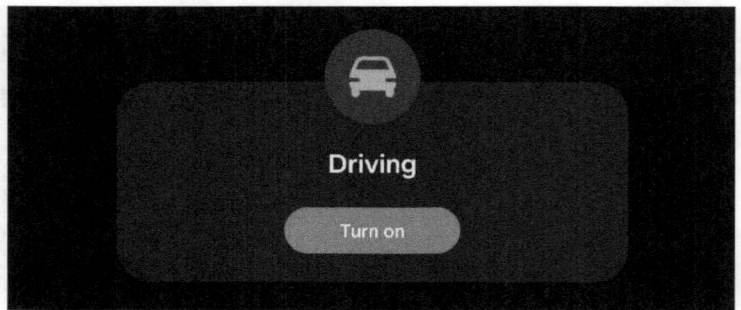

Try this: "Hi, Bixby; turn on Driving mode."

Bixby Routines and SmartThings

Bixby Routines can work with SmartThings, so you can set up a routine that includes controlling your smart home devices and other things. You can access SmartThings when creating a routine in two ways: Modes and Routines and the SmartThings app.

1. **Modes and Routines**: Tap **Routines** at the bottom after opening the **Modes and Routines** app. You can create a new routine or add more actions to an old one. In the "Then" section, click **Add action** if you're adding more actions to a routine, or click **Add what this routine will do** if you created a new one. Find and click on **SmartThings,** or use the search box to find it (*see* **Using Bixby Routines** (page 70)).

2. **SmartThings app**: Tap **Routines** at the bottom after opening the SmartThings app. Click the plus sign to create a routine with the "If" and "Then" sections (*see* **SmartThings** (page 44)).

Section 3: Galaxy AI ✦ – Discover More Smart Features

6: Smart Communication

Good communication really matters – it helps us get along with each other and build stronger relationships. The latest Samsung flagship phones have impressive AI features that break language barriers and make communication easier. This chapter will explore how you can use these smart features to connect with others effectively. These features include:

1. Call Assist
2. Interpreter
3. Chat Assist

Call Assist

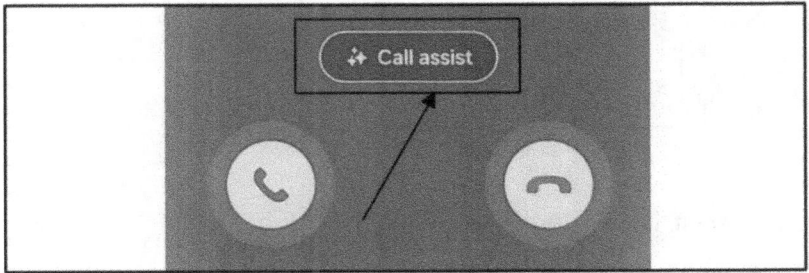

Samsung Call Assist enhances your communication with others by providing real-time assistance during calls.

Follow the steps below to access the Call Assist feature:

- Swipe up from your Home screen to go to your apps list, then find and click on the 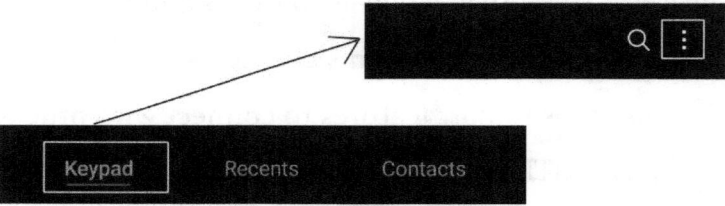 **Phone** app.
    - **Try this**: "Hi, Bixby, open the Phone app."
- Tap **Keypad** at the bottom, and click **More options** at the top right.

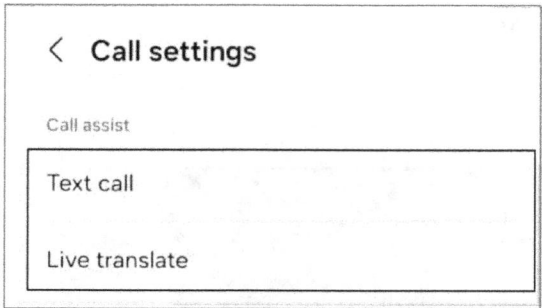

- Next, tap **Settings,** where you'll find Call Assist and other options.

Call Assist is divided into two:

1. Text Call
2. Live Translate

Text Call

This feature is designed to help you receive incoming phone calls without speaking. Instead, you use text to communicate with the other person at the other end of the call. Bixby will read the message aloud to the caller, and whatever the caller says will be converted to text for you to read.

This feature is helpful when you can't take calls but still want to respond to the caller or when you want to initiate a call but can't speak. It may also benefit individuals with difficulty talking on the phone due to speech impairments or hearing loss. Follow the steps below to activate and use this feature:

1. From the 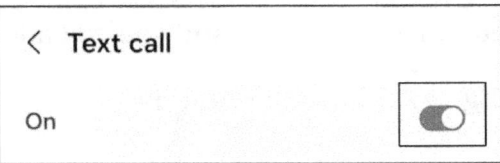 **Phone** app, tap **More options** and click **Settings**.
2. Find and tap **Text call.**
3. Tap the toggle button to enable this feature. You may be prompted to download a language pack.

4. You have three options available: **Language**, **Voice,** and **Quick responses**.

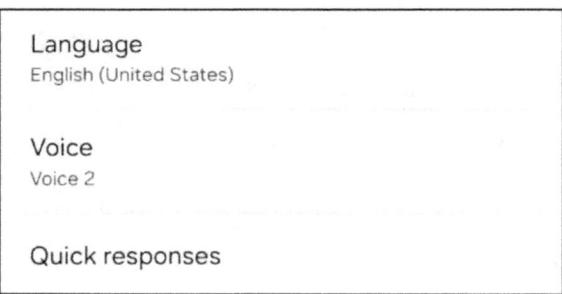

5. Click **Language** to Download the English (United States) language pack or any other language pack.
6. Tap **Voice** and choose "**Voice 1**"(female) or "**Voice 2**"(male). You'll hear a sample of the voice you selected.

   o Tap ➕ **Create custom voice** to make a personalized voice that sounds like you. Just read ten random sentences as prompted and wait for it to process. Once done, the person at the other end would hear a voice similar to yours. I recommend you try it out.

7. When you tap **Quick responses**, you'll see already preloaded quick responses, and you can add more by tapping ➕ **Add quick response**.

Note: The next time you get a phone call, look for the "**Call assist**" option and tap it. Click **Text Call** to answer using this feature. The same "**Call assist**" option appears when you dial a phone number, so tap it and select **Text Call**.

Live Translate

This very cool feature lets you speak to someone in another language. What you say is translated into the person's language, and their response is translated back into yours. Bixby will speak the translated version aloud to you both.

This feature helps overcome language barriers (*see* **Interpreter** (page 97), enabling effective communication between you and others who don't speak your language. Follow the steps below to activate and use this feature:

1. From the 🅒 **Phone** app, tap ⋮ **More options** and click **Settings**.
2. Find and tap **Live translate.**
3. Tap the toggle button to enable this feature. You may be prompted to download a language pack.

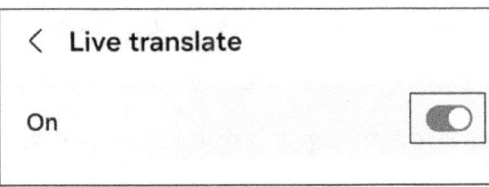

4. You have three options in the "**Me**" and "**other person**" segments. They include:
    - **Language**
    - **Voice**
    - **Mute my voice/Mute other person's voice**

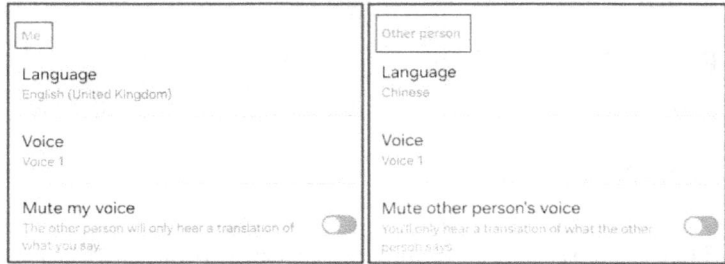

We'll also explore the "**Language and voice presets for each person**" option in the Live translate settings.

5. **Language**: The live translation is done on your phone, so you must download the necessary languages on your device.
    - **Me:** When you tap **Language** in this section, you'll discover it is divided into two parts. The first, "**My language**," has all your downloaded languages – select the one you speak. In the second section, "**Available to download**," tap a language to download and add to the list of downloaded languages.

- **Other person:** When you tap **Language** in this section, you'll discover it's similarly divided into two parts. The first, "**Other person's language**," has all your downloaded languages – select the one the person speaks. In the second section, "**Available to download**," tap a language to download and add to the list of downloaded languages.

6. **Voice**: This option controls the translated voice you and the other person hear during the conversation.

    - **Me:** When you tap **Voice** in this section, you have two options: **Voice 1** (female) and **Voice 2** (male). The voice you select will be what the other person will hear. So, if you're male, choose Voice 2; if you're female, choose the other.

    - **Other person:** When you tap **Voice** in this section, you have two options: **Voice 1** (female) and **Voice 2** (male). The voice you select will be what you'll hear. You can pick any since this will be the default for the other person on the call, and you can't predetermine the caller's identity.

7. **Mute my voice/Mute other person's voice:** This feature lets you control what you and the other person hear during a call.
    - **Me**: Tap the toggle button next to "**Mute my voice**" to ensure the other person hears only your translated voice, not your actual voice (recommended).
    - **Other person**: Tap the toggle button next to "**Mute other person's voice**" to hear only the translated voice of the other person, muting their actual voice (recommended).

8. **Language and voice presets for each person:** When you scroll down, you'll see this option on the "**Live translate**" page. This feature lets you assign a language and a voice to specific contacts.

> Language and voice presets for each person
> Set languages and voices to use for calls with specific phone numbers or contacts.

After clicking "**Language and voice presets for each person,**" follow the step below to assign a language and a voice to a contact.

- On the next page, you'll see the contacts you've added with their assigned language and voice. Tap "**Add number or contact**" to add a new number to the list.

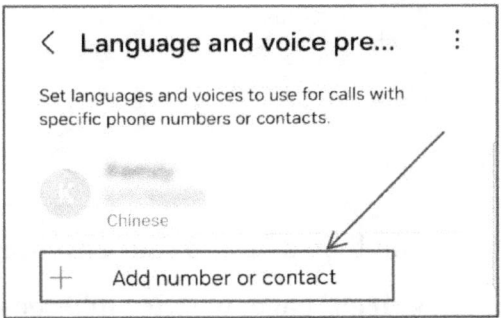

- Find and click on the contact you want to assign a specific language and voice.

- Next, select the language the contact speaks and click **OK**.

- You'll be taken to the "**Languages and voices**" page, where you'll see your selected language and voice settings for "Me" and the selected contact. To modify the voice setting for the selected contact, tap **Voice** and choose either Voice 1 or 2.

- o Tap the back button when done.

Try this: "Hi, Bixby, call James in German."

Note: The next time you get a phone call, look for the "**Call assist**" option and tap it. Click **Live Translate** to answer using this feature. The same "**Call assist**" option appears when you dial a phone number, so tap it and select **Live Translate**.

Interpreter

This function is essential when communicating with someone who is physically near you but speaks a different language. Unlike the Live Translate feature (*see* page 91), which requires you to make a phone call, the Interpreter feature allows for real-time, face-to-face communication.

Accessing Interpreter

There are two ways to access the Interpreter feature. One way is through Quick Settings, and the other is your apps list.

Try this: "Hi, Bixby, open Interpreter."

- **Quick Settings**: Swipe down twice from the top of your screen to display Quick Settings, then find and tap **Interpreter** to open this feature.

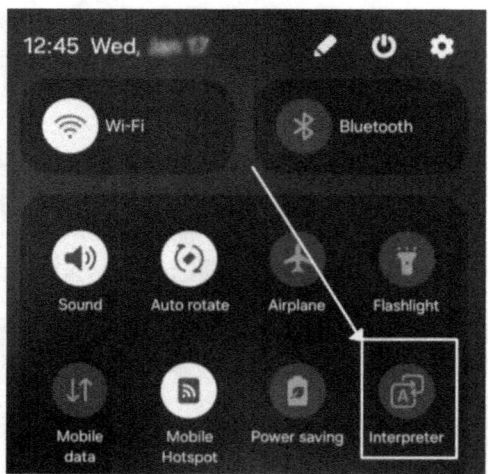

- **Apps list**: You won't find Interpreter in your apps list by default. You'll first need to access Interpreter by going to **Quick Settings** > **Interpreter,** tap

More options at the top-right, then click **Settings** and toggle the "**Show Interpreter on Apps screen**" button.

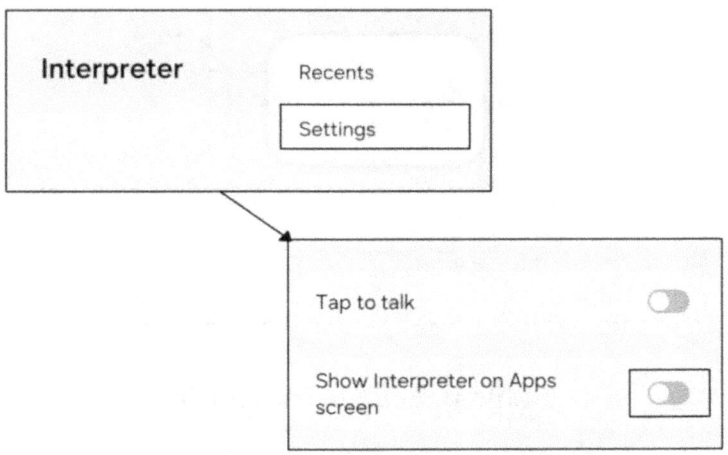

Settings of Interest

You can alter some settings to enhance your experience when using this feature. First, go to **Quick Settings > Interpreter > ⋮ More options**, and tap **Settings** to access the **Interpreter settings**. We'll look at three settings: **Language packs for translation**, **Voice style**, and **Tap to talk**.

Try this: "Hi, Bixby, open Interpreter settings."

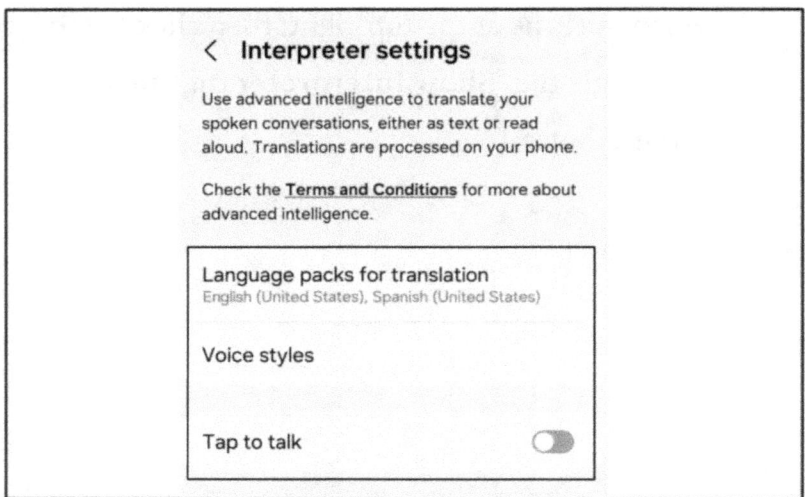

1. **Language packs for translation**: When you tap this option, you'll see two sections, "Installed" and "Available to download." The "Installed" section lists the languages you've already downloaded. The "Available to download" section lets you download more languages when you tap the down arrow next to the language you wish to install and add to your list.

2. **Voice styles:** When you click this option, you'll see all your installed languages. Tap any of them to change their voice style to Voice 1 (female) or Voice 2 (male), and drag the slider to increase or reduce the speech rate. You may want to click English

(United States) and set it to Voice 2 (male) if you're male and an English speaker.

3. **Tap to talk:** When you toggle on this option, you must tap the microphone icon each time to speak. It is helpful for longer sentences that you want to break into segments.

Navigating Interpreter

Let's look at how you can use the Interpreter feature to communicate with someone. Once you've opened Interpreter, there are a few things to know.

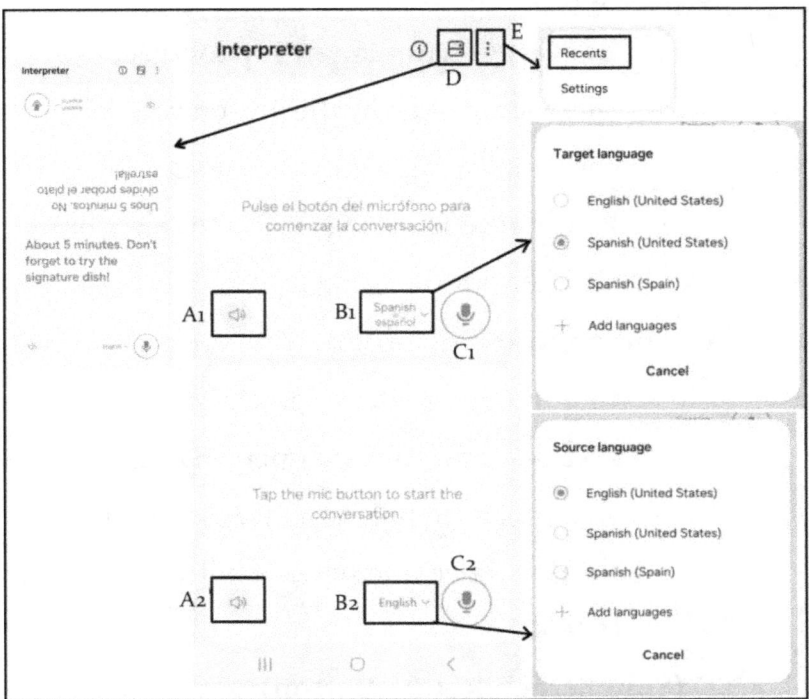

The letters A, B, C, D, and E are labeled in the previous image. We'll use these alphabets to navigate through the Interpreter interface.

- **A (Silent Button):**
    - **A1**: Tap to mute your translated voice so the other person can only read the translated text.
    - **A2**: Tap to mute the other person's translated voice so you can only read the translated text.
- **B (Language Selection):**
    - **B1 (Target Language)**: Tap to select the language spoken by the person you're communicating with.
    - **B2 (Source Language)**: Tap to select or change your own language.
- **C (Microphone Button):**
    - **C1 and C2:** Tap the respective button to speak—C1 for the other person and C2 for you. If "Tap to talk" is active, you must press the respective microphone button each time to talk.
- **D (Split Screen Rotate Button):**

- Tap ⊟ to invert the screen, making it easier for the other person to read the translated text.
- **E (More Options)**:
  - Tap ⋮ to see recent dialogues or to adjust your settings. Tap **Recent** to review past conversations or replay translated dialogues.

Chat Assist

The Samsung keyboard now has new AI assist features to enhance communication. These features include:

1. Chat Translation
2. Style and grammar

Chat Translation

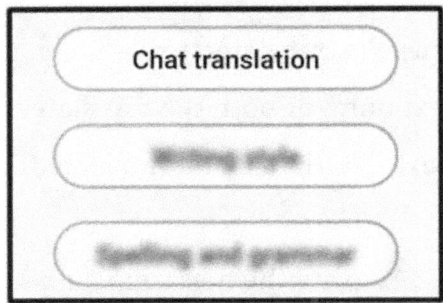

This feature enables real-time translation within messaging apps like the default SMS app, WhatsApp,

Messenger, and others, allowing you to communicate with people who speak different languages. Messages you send are automatically translated into the recipient's language, and incoming messages are translated for you to understand. Follow the steps below to activate and use this feature:

1. Go to your device's **Settings**, tap ✿ **Advanced features** > **Advanced Intelligence** > **Samsung Keyboard** > **Chat translation**, and toggle it on.

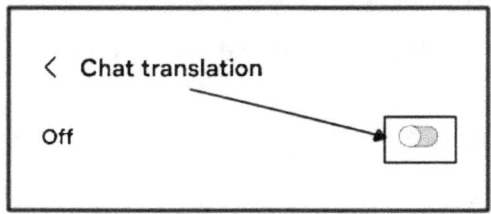

You have three options:
- **Language to show**: Choose whether you want to see only the translated text – "Translated text only" or both the translated and original text – "Original and translated text."
- **Manage apps**: Activate the messaging apps that will use this feature.
- **Language packs for translation**: Download the languages you want to

6: Smart Communication

translate to and from by tapping the down arrow next to them.

2. Next, go to any messaging app to write a text. When the keyboard appears, tap the ✦ Chat Assist button on the left.

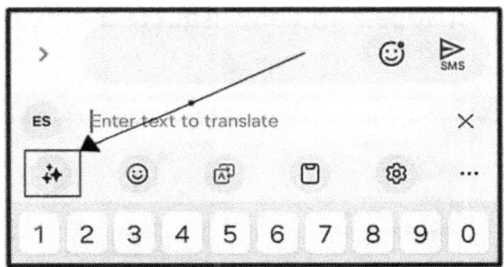

3. You'll find three options: Chat translation, Writing style, and Spelling and grammar. Tap **Chat translation** and select a language by tapping the pop-up at the top. If the pop-up collapses, **Swipe left** on the rectangular bar by the right to open it again.

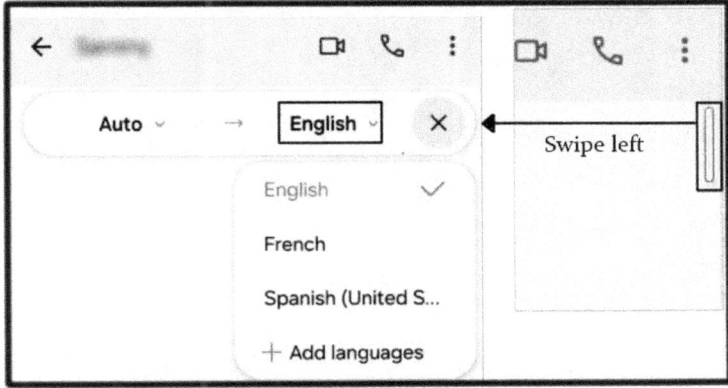

Note: "Auto" automatically identifies the language of the person you're chatting with. Tap **Auto** to choose a language yourself.

4. Write the message you want to send in the text box below the one you normally use. The translated message will appear in your usual text box, then tap send.

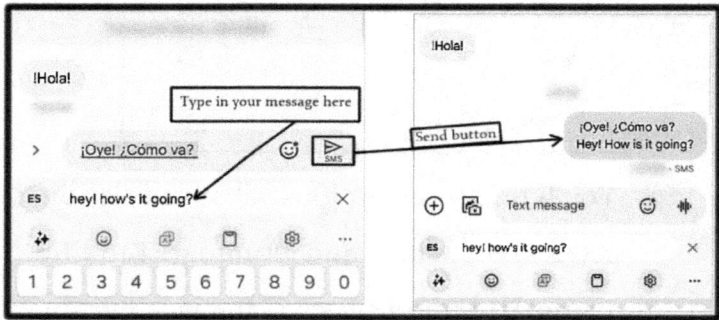

Note: Any messages you receive from the recipient will be translated into your language.

Style and Grammar

This AI assist feature helps enhance your writing style and correct spelling or grammar mistakes. It is divided into two parts: **Writing Style** and **Spelling and Grammar**.

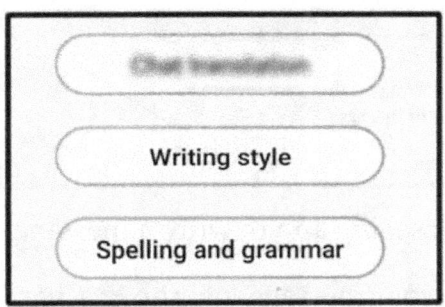

Before we get into these, go to your device's **Settings**, select ⚙ **Advanced features** > **Advanced Intelligence** > **Samsung Keyboard** > **Style and grammar**, and ensure it is toggled on.

1. **Writing Style**: This feature helps improve how your messages sound. It suggests different ways to improve your sentences to make them clearer or more interesting. This feature makes you sound professional or friendly, depending on what you want.

- Go to any messaging app to write a message. After writing your message, tap the ✨ Chat Assist button at the top left of your keyboard.
- Next, tap "**Writing style**." Your text is scanned, and you get some written alternatives, such as casual, polite, and professional.
- Tap "**Copy**" to copy the suggested text or click "**Insert**" to insert it into the text field to replace what you previously wrote.

2. **Spelling and Grammar**: This feature helps correct any spelling or grammar errors in your messages.

- Go to any messaging app to write a message. After writing your message, tap the ✨ Chat

Assist button at the top left of your keyboard.

- Next, tap "**Spelling and grammar**." Your text will be scanned for errors, and you will get a suggestion with all necessary corrections. You can copy or insert the revised text to replace the original.

7: Smart Productivity

Being efficient and productive in your daily life is important, and Galaxy AI offers features to help you achieve that. You can easily find information on anything on your screen, organize, summarize, or translate your written notes, and quickly learn things on the web. Additionally, you can transcribe and summarize your audio recordings to get key points and refresh your memory. These features include:

1. Circle to Search
2. Transcript Assist
3. Note Assist
4. Browsing Assist

Circle to Search

Imagine scrolling through social media and coming across a text or photo that interests you. This feature lets you get more information about that item without leaving the page.

Activating Circle to Search

To use this feature, you first need to activate it. Follow the steps below to turn on this feature.

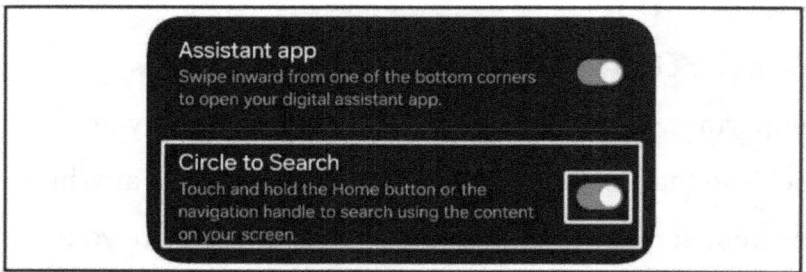

1. Swipe down from the top of any screen on your phone to reveal the Quick Panel. Then, tap ⚙ **Settings** in the top right corner.
2. Scroll down, select **Display** > **Navigation bar**, and toggle on **Circle to search** if turned off.
3. Tap and hold the Navigation bar area of your phone to use this feature. If you're using buttons, tap and hold the home button. Tap and hold the rectangular bar area if you're using swipe gestures.

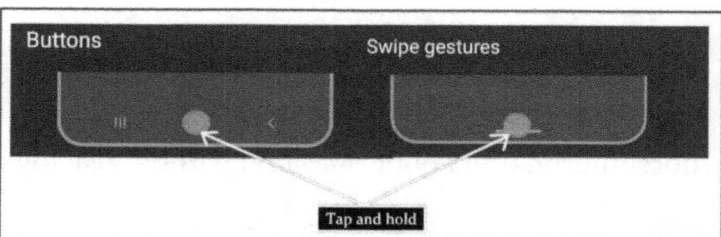

Using Circle to Search

After activating Circle to Search in your device's settings, you're ready to use it. Let's look at three practical examples.

1. Imagine going for a walk with a friend, and something catches your eye (maybe your friend's phone case), and you want one. What do you do?

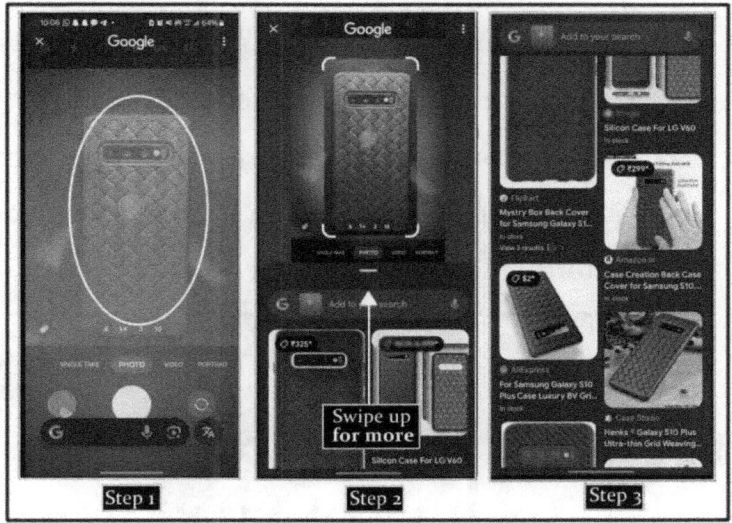

   - First, open the camera app on your phone and point it at the thing of interest.
   - When the item is in focus, instead of taking a picture, tap and hold the home button or the rectangular bar area.
   - Your device should take a screenshot (sort of). If the item is tiny, you can zoom in by

placing your thumb and index fingers on the screen and spreading them apart. Draw a circle around the item on the screen that you previously pointed your camera at (**Step 1**).

- o You will see results related to the item. Swipe up for more results (**Steps 2 & 3**). **Note:** You can use this process to learn about anything you point your camera at, including animals, foods, or plants.

2. Imagine browsing the internet, let's say Amazon, and you come across a non-clickable item you're interested in. What do you do?

- First, tap and hold the home button or the rectangular bar area.
- Next, use your finger or S pen to circle the item you're interested in to select it.
- A result will appear at the bottom, showing more information about the item. Swipe up on it to see additional results.

    Note: You can use this process to get more information on any item shown in an image.

3. Imagine browsing the internet, and you come across a word, phrase, or sentence that intrigues you, and you wish to know more. What do you do?

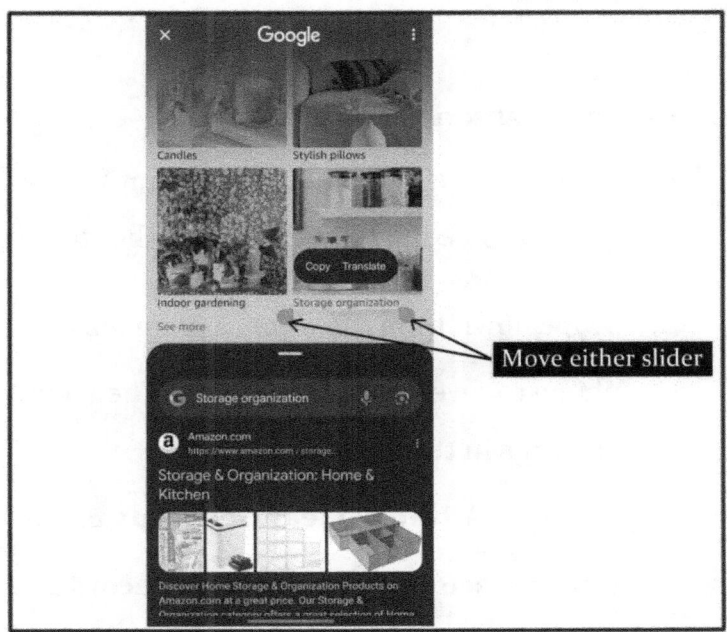

- Again, tap and hold the home button or the rectangular bar area.
- Next, circle the text you want using your finger or S pen.
- The text will be highlighted; you can move either slider to select the specific words or sentences you want.

Transcript Assist

Imagine making a voice recording and converting it into a written transcript and summary. With Transcript Assist, this possibility is now a reality in your 🎙 Voice Recorder app.

Activating Transcript Assist

This feature is active by default. However, there is a setting you may need to turn on. Follow the steps below:

1. Swipe down from the top of any screen on your phone to reveal the Quick Panel. Then, tap ⚙ **Settings** in the top right corner.
2. Select ⚙ **Advanced features** > **Advanced intelligence**, and tap 🎙 **Voice Recorder**. First, select **Transcription language** and choose the

language you speak. Next, tap **Summaries** and ensure it is toggled on. Lastly, tap **Language packs for transcription and translation** to download additional languages.

Using Transcript Assist

The steps below will guide you through how to use this feature and what you should know to enhance your experience:

1. First, open the **Voice Recorder** app from your apps list.
2. You should see a list of your recordings, if any. The ones you've transcribed should have a bunch of words under it.

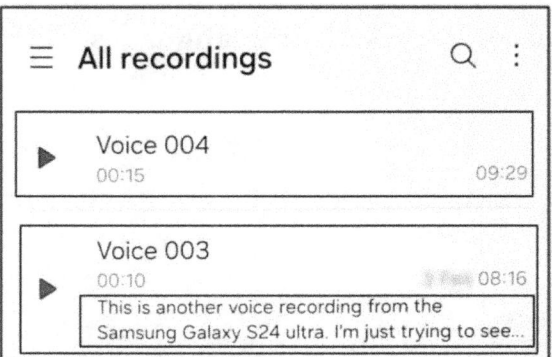

Try this: "Hi, Bixby, translate the second file to French." Note that this command will only work for files you've already transcribed.

3. Click on a recording that has not been transcribed (i.e., one with no bunch of words under it) and tap **Transcribe**. You can also create and save a new recording to transcribe.

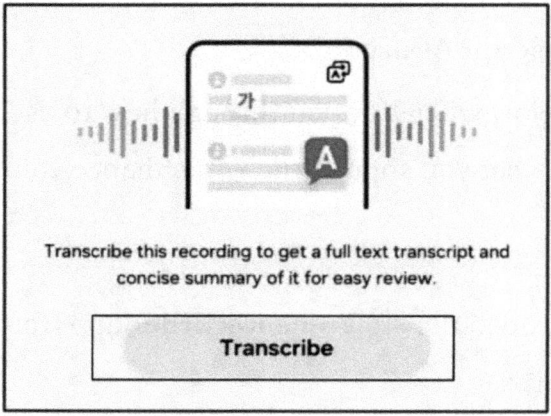

4. After tapping **Transcribe**, choose the language in which the recording was recorded, and tap **Transcribe** again. You can also download more languages by clicking **Add languages.**

7: Smart Productivity 119

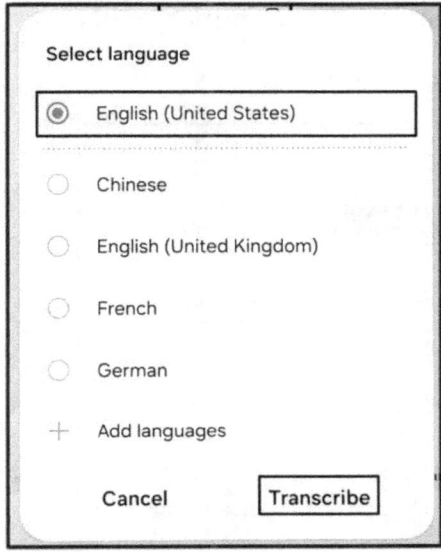

5. On the transcribed page, you'll see the **Transcript** section with everything said in the audio in words, with the speakers differentiated if there's more than one. Tap **Summary** to get a summary of the Transcript.

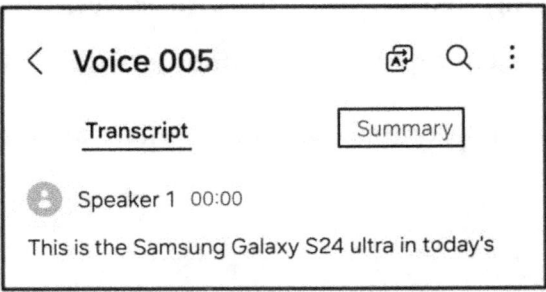

6. Tap the **Translate button**, select a language (Chinese), and click **Translate** to convert your Transcript and Summary into other languages. The

translated Transcript or Summary will appear below the original one.

Note: To translate "Summary," tap **Summary** at the top before clicking the Translate button and following the prompts.

7. You can share and even add the file to your Samsung Notes. First, tap the three vertical dots on the top left, tap **Add to Samsung Notes,** and follow the prompts. Tap the share icon to send

the voice or text file to anywhere or anyone you want.

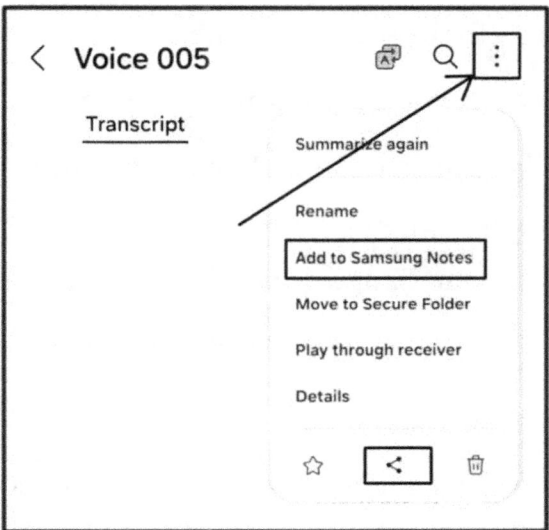

Note Assist

The Samsung Notes app lets you take notes during meetings and lectures or review documents like PDF files. It comes with the Note Assist feature, which is seamlessly integrated to take your experience to another level. This feature allows you to organize, translate, correct spelling errors, and summarize your notes.

Activating Note Assist

To use this feature, you first need to activate it. Follow the steps below to turn on this feature.

1. Swipe down from the top of any screen on your phone to reveal the Quick Panel. Then, tap ⚙ **Settings** in the top right corner.
2. Select ✦ **Advanced features** > **Advanced intelligence**, and tap 🗒 **Samsung Notes**.
3. Next, make sure the feature is on. Then, scroll down and tap **Language packs for translation** to add any language you'd like to translate to.

Using Note Assist

The steps below will guide you through how to use this feature and what you should know to enhance your experience:

1. First, open the 🗒 **Samsung Notes** app from your apps list or ask Bixby to do it.
2. Select a note (hand-written or typed) or create a new one.
3. At the bottom, tap the ✦ Note Assist button and use the slider to select the text you want. You have four options: **Auto format**, **Summarize**, **Correct spelling**, and **Translate**.

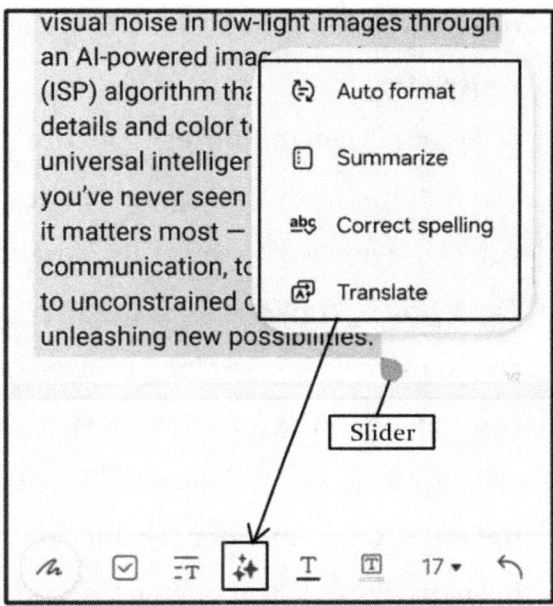

4. **Auto format**: This option helps you organize your notes to make them look more presentable and professional.

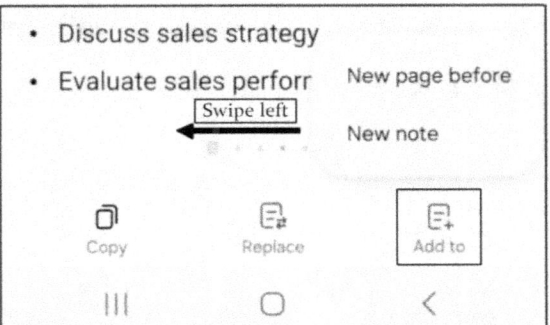

- o When you tap **Auto format** after clicking the ✦ Note Assist button, you can choose from **Headers and bullets** or **Meeting**

notes, whichever best matches the style of your note.
- Selecting "**Meeting notes**" will format the text automatically in that style pattern, and you can swipe left to see other variations.
- Tap **Copy** to copy the formatted text to the clipboard, or click **Replace** to substitute your text with the formatted one. You can also tap "**Add to**" and choose "**New page before**" (Recommended) to add the formatted text to the same note or select "**New note**" to create a new note and add the formatted text to it.

5. **Summarize**: This option lets you summarize your existing notes into shorter versions while retaining all the key points.

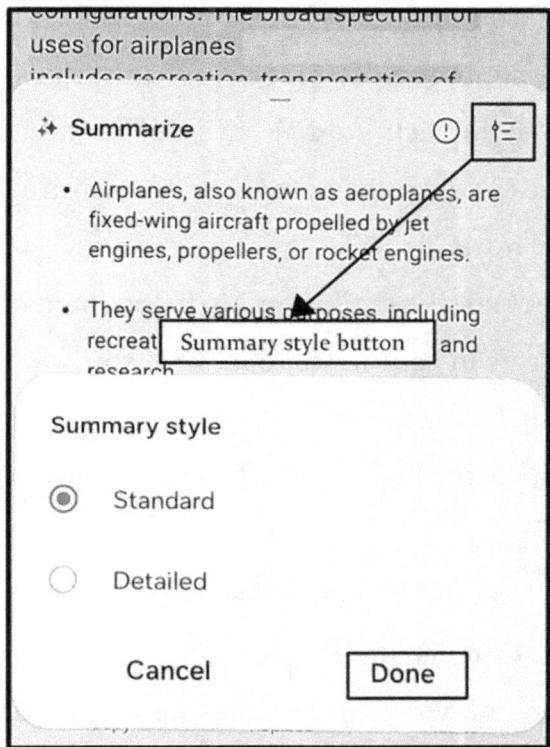

Try this: "Hi, Bixby, summarize this page." Ensure you have the note you want to summarize opened.

- First, tap the ✦ Note Assist button, then tap **Summarize.**
- Your note will be condensed into a shorter version. Tap the **Summary style button** in the right corner and choose **Standard** for a shorter summary or **Detailed** for a longer one, then click **Done**.

- You can "**Copy**" the summarized text to the clipboard, "**Replace**" your existing note with it, or click "**Add to**" to save the text either to a "**New page before**" or in an entirely "**New note**."
6. **Correct spelling**: This option helps you correct all the errors in your notes to make them grammatically correct.

    Try this: "Hi, Bixby, correct the spelling on this page."
    - Tap the ✦ Note Assist button and click **Correct spelling**.
    - Your write-up will be scanned, and the errors will be corrected. You can "**Copy**," "**Replace**," or "**Add to**."

7. **Translate**: This option lets you translate your notes into other languages.

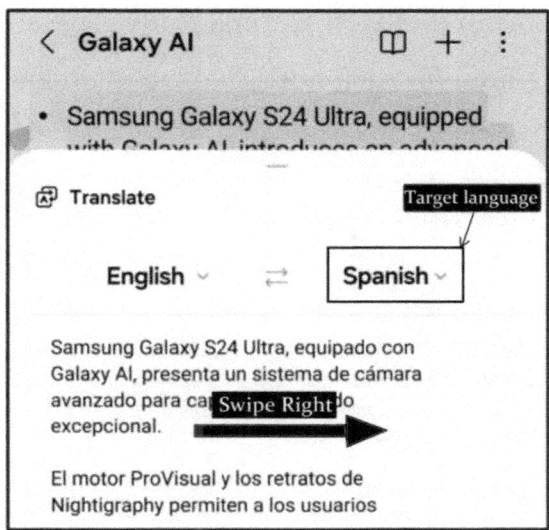

Try this: "Hi, Bixby, translate this page to Spanish."

- Again, tap the ✦ Note Assist button and click **Translate** this time. You may be prompted to select a Target language (i.e., the language you wish to translate into).
- Tap the Target language (**Spanish** in this case) to select another language to translate into and click **Translate**.
- Swipe right on the translated text to see the original. You can "**Copy**," "**Replace**," or "**Add to**."

8. **Generative covers**: Although not directly a part of the Note Assist feature, this option still falls under the same umbrella. It automatically generates a cover photo for each note, making them look clean and easy to tell apart.

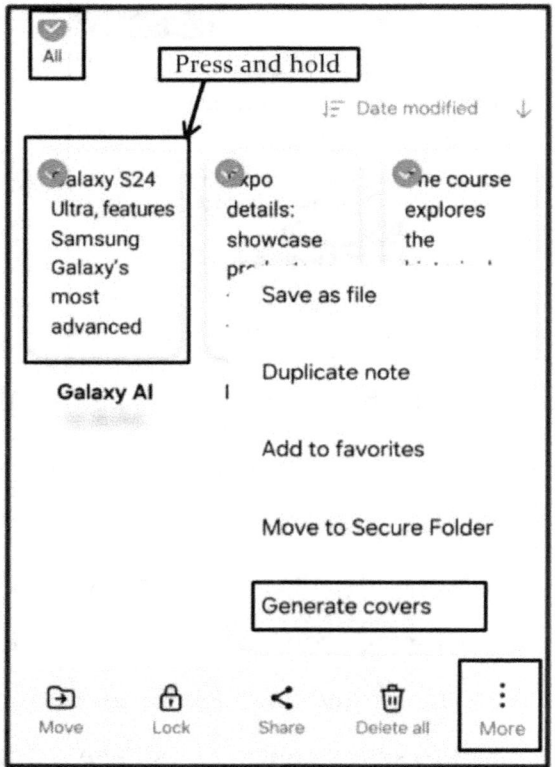

Try this: "Hi, Bixby, create the cover for this note." Ensure you have the note you want to generate cover for opened.

- Navigate to the "**All notes**" page, then press and hold on a note to select it.

- Next, click "**All**" in the left top corner to select all your notes.
- Click ⋮ **More** in the bottom right corner and tap **Generate covers.**
- Once the covers are generated, tap **Done**.

Browsing Assist

The Samsung Internet app now has integrated AI features to elevate your online experience. You can translate a webpage and summarize lengthy pages for quicker reading.

Activating Browsing Assist

You may need to turn on some Browsing Assist features to use them when using the Samsung internet app. Follow the steps below:

1. Swipe down from the top of any screen on your phone to reveal the Quick Panel. Then, tap ⚙ **Settings** in the top right corner.
2. Select ⚙ **Advanced features** > **Advanced intelligence**, and tap 🌐 **Samsung Internet**.
3. You'll find **Summarize** and **Translate** on this page. Tap **Summarize** and toggle it on. You may

be prompted to agree to Samsung account policies. The **Translate** option, on the other hand, is active by default and cannot be toggled on or off. When you tap Translate, you'll see "**Language packs for translation.**" Tap this option and download any language pack you'd like to use.

Using Browsing Assist

The steps below will show you how to use Browsing Assist on the Samsung Internet app:

1. First, open the ◉ Samsung Internet app from your apps list or ask Bixby to do it.
2. Open a web page of your choice, such as an article.
3. Tap the ✦ Browsing Assist button and follow the prompt, if any. You have two options: **Summarize** and **Translate**.

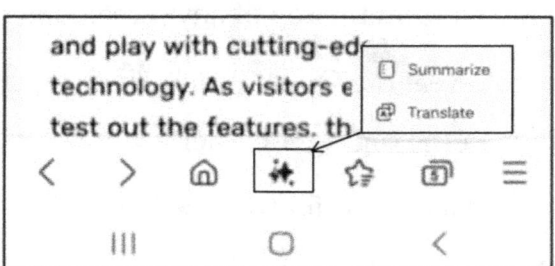

4. **Summarize**: This option summarizes a webpage so you can understand the main points without reading the entire content.

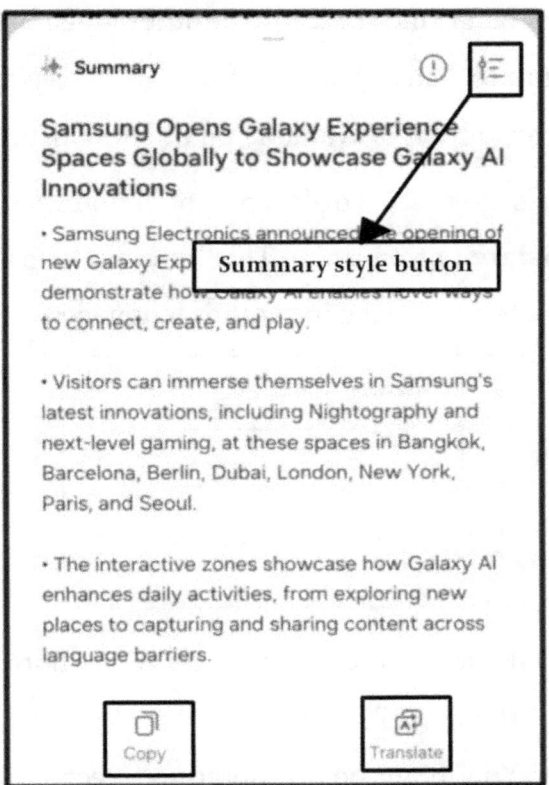

Try this: "Hi, Bixby, summarize this page."

- First, tap the ✦ Browsing Assist button, then tap **Summarize.**
- The webpage you're on will be shortened into a shorter version. Tap the **Summary style button** in the right corner and choose

> **Standard** for a shorter summary or
> **Detailed** for a longer one, then click **Done**.
> - **Copy** the Summarized page to the clipboard or **Translate** it into another language.

5. **Translate**: This option isn't new to the Samsung Internet app. It is available on older Samsung models; go to ☰ **Tools**, and you'll find the **Translate** option there. This function helps translate a webpage into other languages.

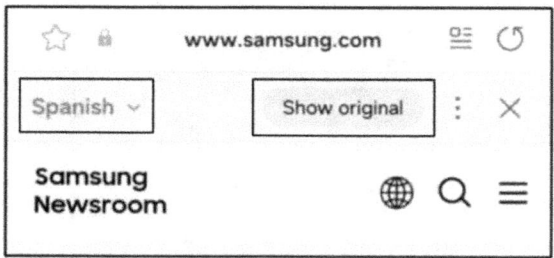

- First, tap the ✦ Browsing Assist button, then tap **Translate.**
- You might be prompted to select a language. To change the language or "Add languages," tap the language (**Spanish,** for instance) at the top left and choose a different language. Click "**Show original**" to see the untranslated content.

8: Smart Personalization

Being creative has never been this easy. Galaxy AI has taken the personalization of your photos and wallpapers to another level. You can now erase, resize, or tilt the subject, an object, or even the overall angle of an image, giving you unique control over how your photos look and feel. This chapter will show you how to use and take advantage of these features. They include:

1. Generative Edit
2. Generative wallpapers
3. Sketch to image
4. Some tips and tricks

Generative Edit

This feature is a new AI tool in the Gallery app that enhances your photo editing abilities. It lets you erase unwanted elements, resize subjects or objects, and straighten images.

Activating Generative Edit

To use this feature, you first need to activate it. Follow the steps below to turn on this feature.

1. Swipe down from the top of any screen on your phone to reveal the Quick Panel. Then, tap ⚙ **Settings** in the top right corner.
2. Select ⚙ **Advanced features** > **Advanced intelligence** > **Photo Editor** and toggle on **Generative edit**.

Using Generative Edit

The steps below will show you how to use Generative Edit to improve your photos in the Gallery app:

1. First, open the Gallery app from your apps list and click on the photo you want to edit.

2. Next, tap ✏️ **Edit** at the bottom to enter edit mode.

3. Click the ✨ **Generative edit** button to start making exciting adjustments to your photo.

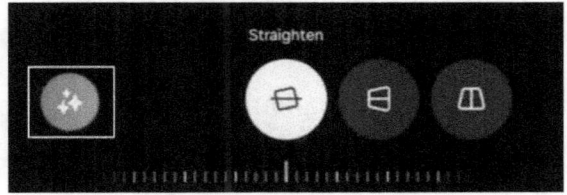

4. Draw a rough circle around the element you want to edit or simply long press on it.
5. When the element is highlighted, there are a couple of things you can do:

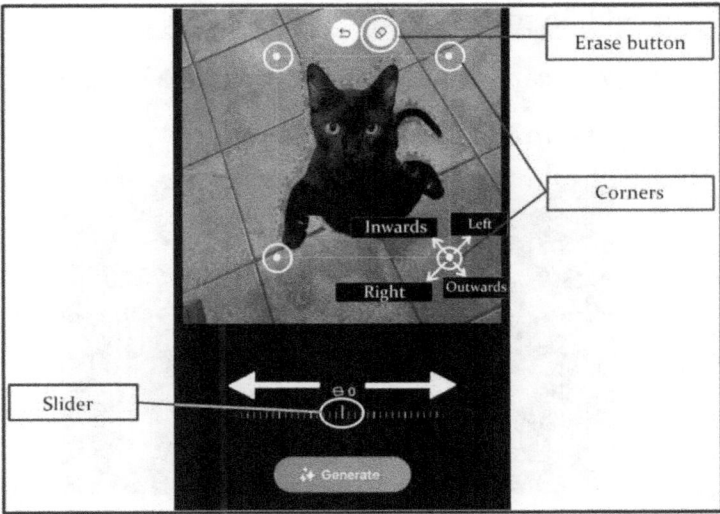

- Press and hold the element's body and drag it to any position you want.
- You can erase the element by tapping the **Erase button**.
- Increase or reduce the element's size by moving the corners **Inwards** or **Outwards**.
- Rotate the element by moving the corners diagonally (**Left** or **Right**).
- Adjust the overall angle of the image by moving the **Slider** left or right up to 15 degrees to keep the element straight or to preference.

6. After completing your edit, tap ✦ **Generate,** and your device will use AI to transform the image and fill in the gaps.
7. You can view the original picture by tapping **View original.** Click **Done.**

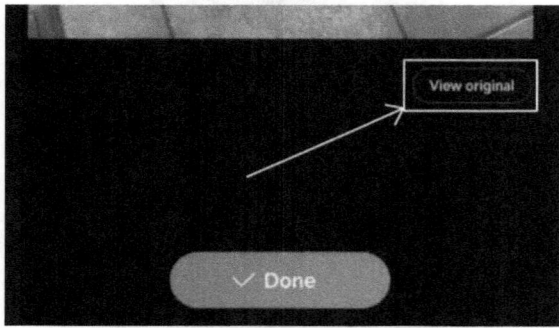

8. Tap **Save as copy** at the top right to save your edited image.

Generative Wallpapers

This feature lets you create wallpapers using AI. Unlike the conventional text prompt where you type what you want, and AI generates it, here you interact with various customizable options. You can adjust colors, select different environments, and tweak other visual elements to craft your unique wallpaper.

Using Generative Wallpapers

The steps below will take you through how to create your very own unique wallpapers:

1. First, swipe down from the top of any screen on your phone to reveal the Quick Panel. Then, tap **Settings** in the top right corner.
2. Scroll down and tap **Wallpaper and style** > **Change wallpapers**. Then, find the **Creative** section and click on it.

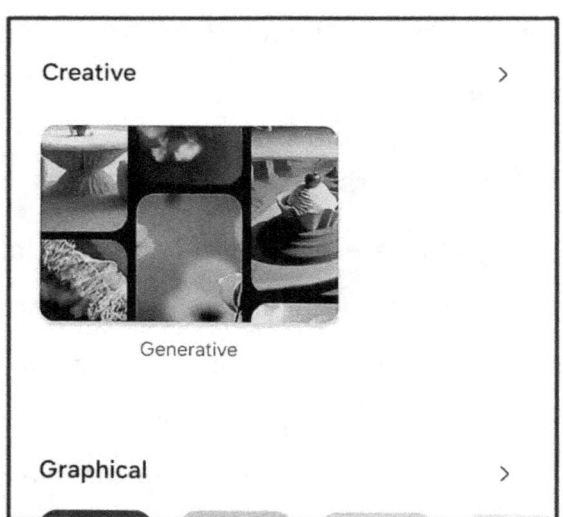

Alternatively, you can access **Wallpaper and style** by pressing and holding on an empty area on your device's Home screen. You'll find **Wallpaper and style** at the bottom left.

3. After clicking the **Creative** section, you'll see different categories under **Create something new.** Choose any option that suits you. For demonstration purposes, let's select **Night**.

4. On the next page, you have a sentence describing your selected image, such as "Abstract **gray distant mountain range** with **ice**." You can tap the highlighted words to change them.

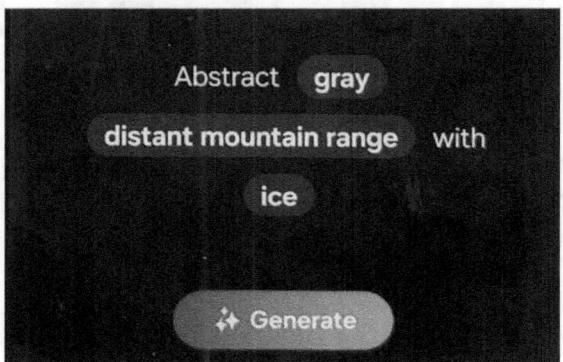

5. When you tap a highlighted word, such as **gray**, you'll be taken to a selection page to choose a different option. You can change **gray** to **gold**, **distant mountain range** to **river**, and **ice** to

clouds. Once you've made your selections, tap **Generate**.

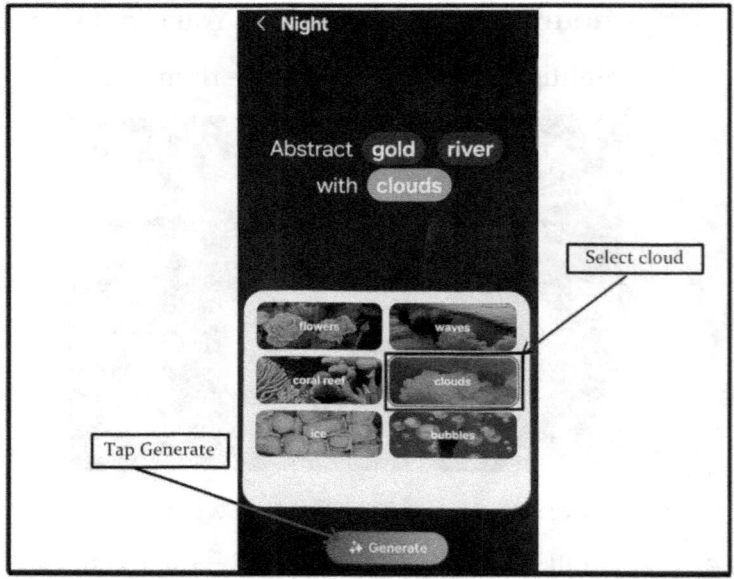

6. A unique wallpaper based on your selections will be generated. **Swipe left** to view other variants. Tap **Set** to choose where to apply the wallpaper. You can set it on your Home and Lock screen or select

just one by checking or unchecking the other. Tap **Next** to confirm your choice, and you're done.

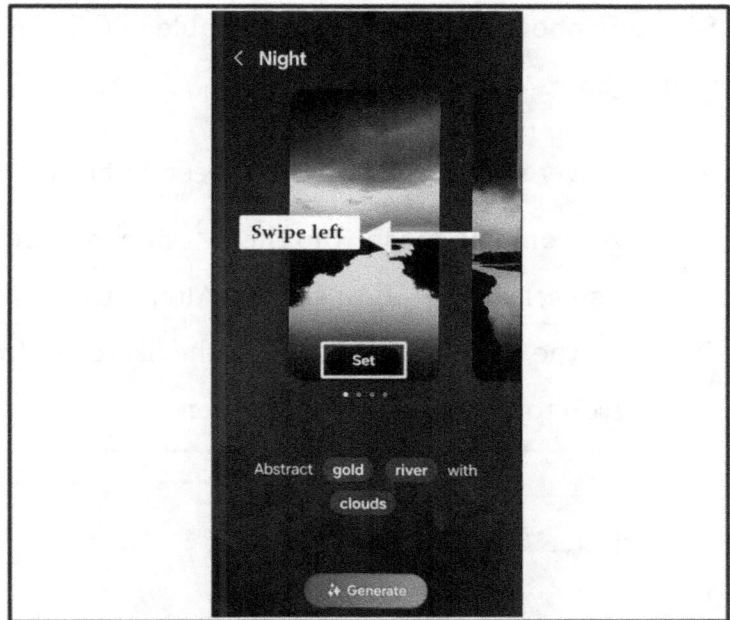

Sketch to Image

This feature was introduced with the Galaxy Z Fold 6 and Flip 6 release and should soon be available on other Samsung flagship devices. With this feature, you can sketch whatever's on your mind and watch AI bring it to life.

Using Sketch to Image

Let's look at how you can use Sketch to Image on your Samsung phone. This feature is available in the Notes and Gallery app.

1. Hover your S pen over any screen to bring up the Air command icon. Tap the ⬤ Air command icon and select **Sketch to Image**. Alternatively, swipe on the **Edge handle** and tap the Sketch to Image icon to open it without an S pen.

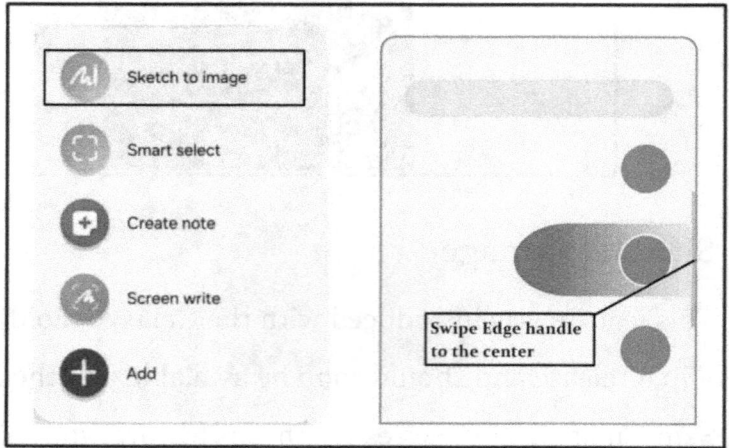

   o You can draw anything you want, and to make it even easier, you can open a picture from your Gallery, the internet, or anywhere. Then, activate the Sketch feature, tap the **Transparency Button** in the top right corner, and adjust it to use the background image as a guide.

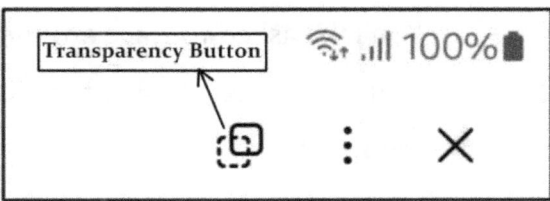

- After you've sketched your drawing, tap the **Styles Button** and choose from one of the five options. Then, tap ✨ **Generate**.

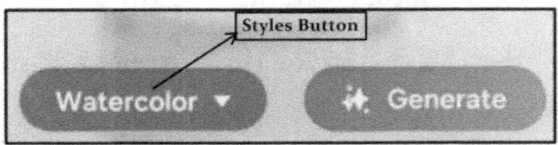

- Swipe left on the generated images to see the different options.

2. Open the 🗒 **Samsung Notes** app from your apps list or ask Bixby to do it. Then, open a new note.

- Click the **Drawing icon** and draw whatever you want, like the sun, as it's easy to draw.

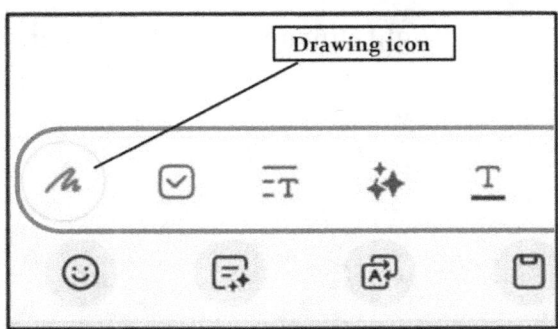

- Next, tap the ✦ Note assist button and click **Sketch to Image**.
- You'll be asked to trace around the drawing, so draw a rough circle or enclosed shape around it.
- Tap ✦ **Generate**. You also have the option to select whatever style you want.

3. Open the Gallery app from your apps list and click on a photo. Next, tap ✎ **Edit** at the bottom, click the ✦Generative Edit button, and tap **Sketch to Image**.

- After you click **Sketch to Image**, draw whatever you want on the image and tap ✦ **Generate**. For example, you could draw a lightbulb and see what results you get.

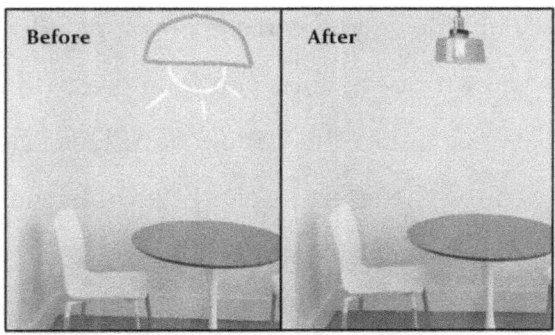

- Swipe left on the image to see other options and click **Save Copy** to save the image.

Some Tips and Tricks

Here are a couple of tips and tricks you might find helpful:

1. Always ask Bixby for help if you're unsure where to find something on your device. Maybe you're looking for a setting, an app, or anything else; Bixby can quickly guide you to the right place.

2. Turn on **Process data only on your device** if you prefer to keep your data more private by processing any AI feature on your device rather than in the cloud. Enable this feature by going to **Settings** > ✦ **Advanced features** > **Advanced Intelligence**, then scroll to the bottom and toggle on **Process data only on your device**.

3. For videos, you can see how any video will look in slow motion when you long press on it while viewing it in the Gallery app. This action allows you to see the effect instantly without making permanent changes.

4. For Images, long press on the subject of an image in your Gallery to isolate it from the background. You can then Copy, Share, or Save it as a sticker. Once you've copied the subject, you can paste it into another image in your Gallery.

About Author

Dylan Blake is a multifaceted creator whose interests span engineering and arts. Since acquiring his first Samsung phone in 2016, Dylan has been fascinated by the rapid evolution of smartphones and the competitive technology landscape. His appreciation for Samsung's innovation and his critical eye for industry advancements fuel his enthusiasm for technology.

Dylan started his writing career with blogs, articles, and poetry and published his first book in 2018. Beyond writing, he is a passionate YouTuber and forex trader, always keen to explore new challenges and opportunities. His diverse pursuits reflect his curious and dynamic nature, constantly pushing the boundaries of what he can learn and teach.

His latest book is born out of a realization that many smartphone users are not fully utilizing the potential of their devices. By focusing on the Advanced Intelligence (AI) features of Samsung phones, Dylan aims to empower users to use technology smarter. He believes that understanding these features can significantly enhance the daily utility of these devices.

Dylan indulges in his passion for cooking and music when he's not exploring the latest in tech. He enjoys crafting new dishes, playing the piano, and singing. Whether creating content for his YouTube channel or analyzing the forex markets, Dylan embraces learning and sharing knowledge. His approach to life reflects a commitment to helping others make the most of technology and beyond.

Index

A

ACCESSIBILITY, 67
Accessing Bixby Modes and Routines, 69
Accessing Interpreter, 98
Activating Browsing Assist, 129
Activating Circle to Search, 112
Activating Note Assist, 121
Activating Transcript Assist, 116
Adding Your Smart Home Devices, 44
Allow personal results, 34
Allow smart home control, 35
Auto format, 123

B

Bedtime routine, 71
Bixby, 23
Bixby country/region, 36
Bixby feedback, 33
Bixby Modes and Routines, 69
Bixby Routines and SmartThings, 82
Bixby Vision, 59
Bixby Vision Features, 62
Bixby Voice, 24
Browsing Assist, 129

C

Call Assist, 87
Caution, 6, 8
Chat Assist, 103
Chat Translation, 103
Circle to Search, 111
Color detector, 68
Communication and Productivity, 40
Correct spelling, 126
Create and Use Quick Commands, 52
Create custom voice, 90

D

DISCOVER, 65
Driving mode, 77

E

Entertainment and Home Management, 42
eSIM, 5

G

Generative covers, 128

H

Health and Fitness, 47

I

Improve voice wake-up accuracy, 31
Improving Bixby Recognition Accuracy, 29
Interpreter, 97
IPX8 rating, 7

L

Language and voice style, 33
Live Translate, 91

M

Maintaining Your Device's Water resistance, 6
My Commands, 55

N

Navigating Interpreter, 101
Note Assist, 121

O

Object identifier, 68
Offline processing, 35
Opening the Bixby Vision app, 61
Optimizing Bixby Settings, 31

P

power off menu, 19

Q

Quick Commands, 50
Quick Commands and SmartThings, 56

R

Recommended, 52
recovery mode, 19
Remapping the Side Key Function, 17
Respond to my voice, 28
Retrain Bixby, 29

S

Samsung account, 11, 24, 32
Scene describer, 68
Setting up Your Galaxy Smartphone, 9
Settings of Interest, 99
Show Bixby on Apps screen, 36
side button, 9, 12, 62
Smart Switch, 15
SmartThings, 44
Sound feedback, 33
Spelling and Grammar, 108
Style and Grammar, 107
Summarize, 124

T

Talk to Bixby without wake-up, 35
TEXT, 64
Text Call, 89
Text reader, 68
Training Bixby to Recognize Your Voice, 27
Transcript Assist, 116
Transfer Data from Your Old Devices, 13
Transferring Data via USB Cable, 14
Transferring Data via Wi-Fi, 16
Translate, 127
TRANSLATE, 62
Travel and Navigation, 49
Try this, 61, 69, 82, 88, 97, 98, 99, 117, 125, 126, 127, 128, 131
Turn off and Restart your Device, 12

U

Unlink device, 36

Updating Bixby Vision, 59
USB-OTG adapter, 14
Use while locked, 34
Using Bixby Modes, 76
Using Bixby Routines, 70
Using Browsing Assist, 130
Using Circle to Search, 113
Using Note Assist, 122
Using Transcript Assist, 117

V

Voice response, 33
Voice wake-up, 25, 32

W

Wake-up phrase, 26
Wake-up sensitivity, 30
Wake-up time routine, 75
WINE, 66
Writing Style, 107